POLITICS OF CHINA'S ENVIRONMENTAL PROTECTION

Problems and Progress

Series on Contemporary China (ISSN: 1793-0847)

Series Editors: Joseph Fewsmith *(Boston University)*
Zheng Yongnian *(East Asian Institute, National University of Singapore)*

---

*Published**

Vol. 6 Water and Development in China: The Political Economy of Shanghai Water Policy
by Seungho Lee

Vol. 7 De Facto Federalism in China: Reforms and Dynamics of Central-Local Relations
by Zheng Yongnian

Vol. 8 China's Elite Politics: Political Transition and Power Balancing
by Bo Zhiyue

Vol. 9 Economic Reform and Cross-Strait Relations: Taiwan and China in the WTO
edited by Julian Chang & Steven M Goldstein

Vol. 10 Discontented Miracle: Growth, Conflict, and Institutional Adaptations in China
edited by Dali L Yang

Vol. 11 China's Surging Economy: Adjusting for More Balanced Development
edited by John Wong & Wei Liu

Vol. 12 Tobacco Control Policy Analysis in China: Economics and Health
edited by Teh-Wei Hu

Vol. 13 China's Science and Technology Sector and the Forces of Globalisation
edited by Elspeth Thomson & Jon Sigurdson

Vol. 14 Migration and Social Protection in China
edited by Ingrid Nielsen & Russell Smyth

Vol. 15 China's Reforms at 30: Challenges and Prospects
edited by Dali L Yang & Litao Zhao

Vol. 16 Political Booms: Local Money and Power in Taiwan, East China, Thailand, and the Philippines
by Lynn T White

Vol. 17 Politics of China's Environmental Protection: Problems and Progress
by Chen Gang

*To view the complete list of the published volumes in the series, please visit:
http://www.worldscibooks.com/series/scc_series.shtml

Series on Contemporary China – Vol. 17

POLITICS OF CHINA'S ENVIRONMENTAL PROTECTION

Problems and Progress

by

Chen Gang
East Asian Institute
National University of Singapore, Singapore

World Scientific

NEW JERSEY • LONDON • SINGAPORE • BEIJING • SHANGHAI • HONG KONG • TAIPEI • CHENNAI

Published by

World Scientific Publishing Co. Pte. Ltd.

5 Toh Tuck Link, Singapore 596224

USA office: 27 Warren Street, Suite 401-402, Hackensack, NJ 07601

UK office: 57 Shelton Street, Covent Garden, London WC2H 9HE

British Library Cataloguing-in-Publication Data
A catalogue record for this book is available from the British Library.

Series on Contemporary China — Vol. 17
POLITICS OF CHINA'S ENVIRONMENTAL PROTECTION
Problems and Progress

ISBN-13 978-981-283-869-8
ISBN-10 981-283-869-4

Typeset by Stallion Press
Email: enquiries@stallionpress.com

Printed and bound in Great Britain by
Marston Book Services Limited, Oxfordshire

To Lixi and Xinglin

Contents

Preface xiii

Acknowledgments xvii

Introduction xix

Chapter 1. Environmental Crises in China 1

China on the Brink of Water Crisis 1

Severe Water Pollution 1

Water Shortage 7

Air Pollution 8

Climate Change 10

Other Problems 13

Chapter 2. Bureaucratic Structures on Environmental Protection 16

Interaction among the Party, the Administration, and the Legislature 16

Central-Level Administrative System on Environmental Protection 18

Weakness in Local Environmental Protection Departments 22

Clash between GDP Growth and Pollution Control 26

Environmental Impact of China's WTO Membership 29
Chapter 3. Environmental Legal Systems in China 33
Basic Environmental Law 33
Specialized Environmental Laws and Regulations . . 35
Environmental Dispute Settlement 36
Statutory and Enforcement Flaws 38
Chapter 4. Environmental NGOs and Emerging Civil Society 41
The Development of Environmental NGOs in China 42
Greening of Media and Internet 45
Political Constraints and Challenges Facing Chinese ENGOs 48
Policy Constraints 48
Weak Capacity Building and Internal Management 50
Chapter 5. China's Environmental Diplomacy and Soft Power 53
China's Environmental Diplomacy 54
Soft Power and Institutional Participation 61
Institution Participation Enhances Soft Power . 62
China's Active Participation in International Institutions 64
Chapter 6. Case Study: Songhua River Toxic Spill Crisis 66
Review of the Songhua River Pollution Crisis 67
Who Should Be Responsible? 69
Governance Problems Reflected 73
Chapter 7. Case Study: Taihu Algae Crisis 76
Failure of 16-Year Effort to Clean-Up Taihu Lake . . 77
Crazy for Gross Domestic Product (GDP) 78
Wu Lihong's Tragedy: Failed Crusade Against Pollution 82
Possible Solutions 85
Chapter 8. Case Study: Beijing Olympics and Air Quality 88
Smog Casts Shadow over "Green Olympics" 88

Questionable "Blue Sky" Achievements 91
Sources of Air Pollution 94
Contingency Measures and Solutions 96
Chapter 9. China's Climate Change Politics 103
China's Performance in Pre-Kyoto Climate Negotiations 104
Start of International Climate Talks under the UN Framework 104
China's Negotiation Stance 105
China's Gains from Kyoto Protocol 107
Diplomatic Scores 108
Support from Developing Countries 108
Praise from Some Industrialized Countries . 108
Soft Power Enhancement through Institutional Participation 109
Economic Gains 110
Huge Lucre from CDM Projects 110
Optimizing Economic Structure and Accelerating Technology Upgrade . . . 112
Easing Energy Security Anxiety 113
Improving Domestic Governance 114
"Development First" and China's Inter-Agency Mechanism on Climate Change 115
National Climate Change Program's Focus on Development 115
Inter-Agency Mechanism on Climate Change . 117
Domestic Pressure and China's Initiative 119
China's Future Stance in Post-Kyoto Negotiations . . 121
Voluntary Targets Instead of Absolute Emission Obligations 121
Continued Solidarity with Developing Countries 123
More Attention to Regional Mechanism 125
Chapter 10. How China Manages and Benefits from CDM Projects 127
China's Huge Potential in CDM Projects 128
CDM and Foreign Investment Inflow 128

Chinese Government's Revenue from CDM Projects 130
Two CDM Cases 130
China's Share in Global Carbon Market 131
The Largest Piece of Cake 132
Rising Value of Emission Reductions 134
Government's Endeavor to Boost CDM Projects 136
Procedures of China-Based CDM Projects 138
Project Identification 139
Project Design 139
Host Country's Approval 139
Project Validation 140
Project Registration 140
Project Implementation, Supervision, and Reporting 140
Verification and Certification of CERs 140
Issuance of CERs 140
Challenges and Obstacles 141

Chapter 11. Reforms and New Policies 144

New Policies and Initiatives 144
Suspension of Approval of New Industrial Projects along Major River Systems (*liuyu xianpi*) 145
Electricity Generated by Renewable Energy to Be Purchased Preferentially 145
"Green-Credit Policy" (*luse xindai zhengce*) . . . 146
Establishing Leading Group on Energy Saving and Emission Cutting 147
Using Green Indexes to Measure Officials' Performances 148
Establishment of Green Fund 149
Eco-City Project 149
Discussion on Levy of Environmental Tax . . . 150
Establishing Pilot Zone on Resource Efficiency and Environmental Protection 151

Promoting "Green Insurance" 152
Elevating SEPA to Full-Fledged Ministry — Ministry of Environmental Protection 153
A Long Way to Go Green 154
Bibliography 159
Books and Book Chapters 159
Journal Articles 161
News Media 162
Other References 162
Websites 163
Index 165

Preface

The environmental conservation has made it to the highest platform in the ruling Chinese Communist Party (CCP). General Secretary Hu Jintao's report to the 17th party congress floated the notion of "conservation culture" for the first time, in the context of the need to "build a comprehensive, well-off society". The high environmental cost accompanying economic growth tops all other problems, said Hu, in a clear signal that such issues, once marginalized by senior decision-makers, have become a new policy focus.

As widely reported, the country's double-digit economic growth rates have led to serious environmental problems, including worsening air conditions in the cities, rocketing greenhouse-gas emissions, serious water shortage, and pollution as well as desertification and loss of biodiversity. Environmental improvement, however, is not only related to the speed of economic development, but also closely links with the country's political structure and bureaucratic systems. In other words, ecological protection is not only an issue of environmental engineering, technology, and economics, but also a tough political task that has to resort to the improvement of governance and policy adjustment. China's environmental pollution is to a larger extent a political issue concerning ineffective governance rather than the one caused by insufficient technology or fund.

Tackling environmental issues is important not only for its own sake but also for the CCP's political survival as well as the country's soft power in the international arena. The experience of Eastern Europe and the Soviet Union shows that growing environmental discontent often served as a catalyst for broader opposition to communist regimes. In China, protests about environmental issues have been growing by 30% annually, and as a result, the Chinese government has been using another catchphrase "coordinated development" (*xietiao fazhan*) to show its increasing attention to sustainable development, reduction of energy consumption, and protection of the environment.

It is unfair to criticize the Chinese government for not making any substantial achievement in ecological protection during the past three decades of reform and opening-up. Starting from nil in early 1970s, China has successfully built up the national regulatory framework of environmental protection with increasing number of employees and a complicated legal system in this area containing specialized environmental laws, instruments, and regulations. More importantly, after new leadership cored by Hu Jintao and Wen Jiabao gradually took paramount power after 2002, the ruling party has enacted a series of fresh administrative initiatives to address the urgent environmental degradation. Such effort picked up its momentum in 2007 after the country experienced a few shocking environmental crises and was expected to soon replace the United States to become the world's largest greenhouse gas emitter.

As the dazzling economic and social changes in China have imposed substantial impact upon the quality of environmental governance, it is time to review the problems and progress in politics of China's environmental protection. The book tries to answer the following questions: what are the major environmental problems faced by China nowadays? What are China's basic political structure and legal system involved in environmental protection and what are the interrelationship and power distribution among major players? What are the fundamental institutional factors that hamper the Chinese government/party from being successful and efficient in controlling pollution? How has the environmental issue influenced China's foreign policy? How has the participation in international politics influenced China's domestic environmental governance? What are the new administrative initiatives taken by the government to address environmental pollution and will these actions be effective? In answering the above questions, here emerges the fundamental question: after Chinese leaders have changed their previous uninterested stance and become extremely sensitive to environmental issues, will pollution and other environmental problems be solved much more effectively and no longer pose a threat to the CCP governance?

This book examines two environmental crises in the Hu-Wen era, trying to test whether China's environmental governance has been greatly improved with the establishment of mature mechanisms concerning crisis management, accountability and inter-agency coordination. China's effort to improve its capital's air quality before the 2008 Beijing Olympic Games provides another meaningful perspective for people to understand that environmental protection in China on many occasions has become a political show that demands enormous fund and personnel investment and a scurry of policy initiatives. All three cases reflected that high-profile governmental actions may have cheerful effect in controlling pollution in the short run, but the entrenched institutional factors will prevent the problems from being fundamentally solved or similar events from reoccurring.

China's climate change policy is included in this book not only because global warming has come into the limelight of international community in recent years, but also because it offers a unique dimension to analyze the country's environmental diplomacy and domestic bureaucratic structure on emission cutting and related energy issue. The book details on how the government manages emission reduction projects within its territory and becomes the world's largest beneficiary from the Kyoto Protocol, hoping to provide guidance for investors interested in China's burgeoning carbon market and reaching a conclusion that the Chinese government can become very efficient in protecting the environment when certain economic incentives are offered.

The book finally talks about new policies and reform measures in the green area taken by the government since 2007 to handle the hot potato, arguing that some of them may be quite effective in the long run as long as they alter institutional factors and growth-first mindset that obstruct green effort.

This book may be of interest to the following readers: China watchers, environmentalists, social scientists concerned with environmental governance and ecological economics, students in courses dealing with contemporary China and public policy of environmental protection, as well as business people with the intention to tap market potential of China's environmental industry.

Acknowledgments

The book was drafted and completed during my work as a full-time research fellow at the East Asian Institute (EAI), National University of Singapore. The EAI's mission is to promote academic and policy-oriented research on contemporary China and other East Asian economies. Without a specialized library with a good collection of English and Chinese materials on China and other research facilities provided by the EAI, the completion of the manuscript would have become impossible. My deepest thanks go to the institute as a whole for the extensive support at many stages of my work.

More specifically, I am pleased to acknowledge my debt to my EAI colleagues who supported me and gave me valuable guidance, advices, and help during the process of data collection and manuscript preparation. These colleagues include Professor Wang Gungwu, Professor John Wong, Professor Zheng Yongnian, Professor Yang Dali, Dr Bo Zhiyue, Dr Heng Siam-Heng, Dr Huang Jing, Dr Lam Peng Er, Dr Yang Mu, Dr Lim Tai Wei, Dr Tong Yueting, and Dr Zhao Litao. I am grateful to all of the people mentioned above, and the regular exchanges with them helped me to study China's environmental governance issue from various perspectives.

I would also like to express my gratitude to the following scholars and experts for their assistance. During his three-month stay at the EAI as a visiting senior research fellow, Dr Yang Guobin, Associate Professor with the Barnard College, Columbia University, gave me special help on the topic of China's environmental NGOs and emerging civil society by offering his insightful analysis and recommending related journal papers for reference. Dr Tang Shiping and Dr Li Mingjiang, two China experts at the S. Rajaratnam School of International Studies in Singapore, have for long gave me persistent encouragement in academic work and penetrating views on China's foreign policy and soft power issue. Same appreciation goes to Professor Qin Yaqing and Professor Zhu Liqun at the China Foreign Affairs University, Dr Wang Zhengxu and Dr Lai Hongyi at the School of Contemporary Chinese Studies, the University of Nottingham, Mr Huang Langhui with China's National Bureau of Statistics, Professor Wu Yanrui at the Department of Economics, University of Western Australia, and Professor Wang Tianfu at the Department of Sociology, Tsinghua University.

I would like to thank World Scientific Publishing Company, one of the leading scientific publishers in the world, for the excellent editorial work. I am particularly grateful to Professor Kok-Khoo Phua, Chairman of the World Scientific Group, and Mr Max Phua, executive director of the group, for their intensive interest in China's environment issues and enthusiastic support for the publication of this book. Thanks also go to some of its staff including Ms Chua Kim Lee, Ms Chen Yihan, Ms Tai Xiang, and Ms Zhang Hui Mei for their indispensable assistance in producing this work.

Introduction

Since China deepened its reform and opening-up through introducing the expression "socialist market economy with Chinese characteristics" and thus unleashing market forces in the early 1990s, its gross domestic product (GDP), especially in the coastal areas, has kept a high-speed growth momentum for over 15 years. This has imposed huge pressures upon the country's already worsened environment and scant resources. Its mounting environmental problems have caught intensive attention from the Chinese government, domestic public as well as the international community. Water and air pollution, desertification, and other environmental degradation are now so severe that they not only threaten the sustainable development of the economy, people's health, and ordinary life, but also pose an acute political challenge to the governance of the Chinese Communist Party (CCP).

The country is home to 16 of the world's 20 most polluted cities,[1] and only 1% of the country's 560 million city dwellers breathe air considered safe by the European Union.[2] As latest scientific findings describe the connection between climate change and human activity with more accuracy,

[1]Elizabeth C. Economy, "The Great Leap Backward?", *Foreign Affairs*, Vol. 86, Issue 5, (2007), p. 40.

[2]Joseph Kahn and Jim Yardley, "As China Roars, Pollution Reaches Deadly Extremes", *The New York Times* (26, August 2007).

China, estimated by the International Energy Agency (IEA) to become the world's largest emitter of greenhouse gases in 2007, is now one of the countries under closer international scrutiny. China is facing the serious problem of access to clean water. Although China holds the fourth largest freshwater resources in the world (after Brazil, Russia, and Canada), skyrocketing demand, overuse, inefficiencies, pollution, and unequal distribution have produced a situation in which two-thirds of China's approximately 660 cities have less water than they need and 110 of them suffer severe shortages.[3] China is among the world's largest desert nations. A Ministry of Science and Technology task force says desertification costs China about US$ 2–3 billion annually, while 800 kilometers of railway and thousands of kilometers of roads are blocked by sedimentation.[4] An estimated 110 million people suffer firsthand from the impacts of desertification and, by official reports, another 2500 square kilometers turns to desert each year.[5]

There is no doubt that China's high-speed economic growth over the last 30 years has had some positive impacts on the environment, given the fact that technology improvement and fund abundance are playing a conducive role in optimizing resources. Energy efficiency has improved drastically — almost three times better utilization of energy resources in 2000–2002 compared to 1978.[6] As a result of the changing industrial structure, the application of cleaner and more energy-efficient technologies, and pollution control efforts, ambient concentrations of particulate matter (PM) and sulfur dioxide (SO_2) in cities have gradually decreased over the last 25 years.[7] Implementation of environmental pollution control policies — particularly command-and-control measures, and also economic and voluntary measures — have contributed substantially to leveling off or even reducing pollution loads, particularly in certain targeted industrial sectors.[8]

[3] *Ibid.*
[4] Ron Gluckman, "Beijing's Desert Storm", http://www.gluckman.com/ChinaDesert.html.
[5] *Ibid.*
[6] The World Bank and SEPA, "Cost of Pollution in China: Economic Estimates of Physical Damages", The World Bank, Publications and Reports, March 2007, p. xi, http://siteresources.worldbank.org/INTEAPREGTOPENVIRONMENT/Resources/China_Cost_of_Pollution.pdf.
[7] *Ibid.*
[8] *Ibid.*

Table 0.1 China's Priority Environmental Problems

- Water pollution, especially contamination by organic waste
- Water shortages, particularly in northern China
- Urban air pollution, as measured by particulates and SO_2
- Hazardous and toxic solid waste in urban areas
- Soil erosion
- Loss of forests and grasslands
- Loss of species and habitats, especially wetland

Source: Xiaoying Ma and Leonard Ortolano, *Environmental Regulation in China: Institutions, Enforcement, and Compliance* (Oxford: Rowman & Littlefield Publishers, 2000), p. 2.

At the same time, however, the economic miracle is accompanied by mounting environmental woes and the country is fast becoming one of the leading polluters in the world. China's priority environmental problems include water pollution, water shortages, air pollution, hazardous waste, soil erosion, loss of forests and grasslands, and loss of species and habitat (Table 0.1).

The year 2007 witnessed some very serious environmental incidents in China. In May and June 2007, a severe algae outbreak in the Taihu Lake, China's third largest freshwater lake, rendered tap water foul-smelling and undrinkable over the course of a week for half of the 2.3 million residents in Wuxi, a wealthy city in eastern China's Jiangsu Province. This worst-ever water crisis in Wuxi immediately caused social panic, touching off local residents' scare buying of bottled water in the supermarkets and even class suspension of some local colleges.

High levels of nitrogen and phosphorus are believed to be major causes of algae blooms, which develop in water that is rich in nutrients. Apart from Taihu Lake, blue-green algae outbreaks have been reported in Chaohu Lake and southwestern Dianchi Lake since May 2007, and have threatened the local tap water supply. The blue-green algae outbreak in large lakes was only one symbol reflecting China's rapidly deteriorating environment and ecosystems in recent years, when the country was turning itself into a "world workshop" with economy surging at double-digit growth rate per annum. In May and June 2007, several hundred people staged a peaceful demonstration against a planned chemical plant named Tenglong Aromatic PX Company, in southern China's Xiamen City.

The work was suspended by the government after residents sent about one million mobile phone short messages in protest against possible health dangers.

Water crisis, air pollution, climate change as well as growing protests triggered by environmental problems are testing Chinese leaders' wisdom to balance economic growth with environmental problems. If the Chinese Communist Party wants to consolidate its rule, it has to address environmental issues properly and timely. Examples from Eastern Europe and the Soviet Union show that growing environmental discontent often served as a catalyst for broader opposition to the communist regime. A notorious accident at the Chernobyl nuclear power plant in Ukraine in 1986, for example, undermined faith in the Soviet authorities and helped accelerate the breakup of the Soviet Union.

Environmental conservation has made it to the highest platform in the CCP, whose General Secretary Hu Jintao's political report to the 17th Party Congress in 2007 floated the notion of "conservation culture" (*shengtai wenming*) for the first time, in the context of the need to "build a comprehensive, well-off society".

Environmental pollution is a global issue that most fast-growing economies have to face, relating to many factors such as economic structure, technological level, political systems, governance capacity, institutional building as well as public awareness and social participation. In China, however, the increasing environmental crisis is more likely a result of its unique political and economic institutions, only showing the weak capacity of the government to handle those "growing pains". This book, putting an emphasis on China's current environmental policy, regulations, related enforcement issue, and the bureaucratic politics concerning environmental protection, shows governance improvement is a better solution to China's pollution, rather than simply enhancing fund and technology investment.

In recent years, China's deteriorating environment and related governance problem have drawn growing attention from China watchers and other researchers outside the country. In January 1998, a conference entitled "The Chinese Environment" was convened by *The China Quarterly* at the School of Oriental and African Studies, University of London. The discussion papers for the conference were later compiled into *Managing the Chinese Environment*, a book edited by Richard Louis Edmonds and published by the Oxford University Press.[9] This book covered a wide range of China's environmental problems, including land resources, water pollution and shortage, biodiversity and energy use. In his chapter "The Environmental Legacy of Imperial China", Mark Elvin argued in the book that environmental problems are not new to China, which has pushed its growth beyond sustainable limits on many occasions prior to the modern era. In her contribution to the book, Abigail R. Jahiel said that there are problems of insufficient authority and a lack of coordination in China's bureaucracy involved in environmental protection, while the overall management ethos in China remains one in which sustaining economic growth is more important than sustaining the environment.[10] In another book *Environmental Regulations in China: Institutions, Enforcement, and Compliance*, the authors emphasized the institutional factors influencing the environmental protection in China, giving a much more detailed picture about China's legal and administrative system concerning ecological conservation.[11] The book pays particular attention to the customs and informal codes of behavior that govern relationships between enterprises and regulators and to the post-1978 economic and administrative reforms that are moving China toward a market economy. The authors interestingly argued that *guanxi*, the Chinese word frequently translated as "social connections", can be used by entrepreneurs to avoid compliance of environmental regulations or can assist in enhancing environmental protection in both ways.[12] The book, *China's Past, China's Future: Energy, Food, Environment*, argued that the gradual lowering of subsidies on energy use and deregulation of coal prices have helped China to reduce its energy intensity remarkably.[13]

In cooperation with the Chinese government, the World Bank has since 2003 put together a joint Chinese and international expert team that

---

[9]Richard Louis Edmonds (ed.), *Managing the Chinese Environment* (Oxford: Oxford University Press, 1998).

[10]Abigail R. Jahiel, "The Organization of Environmental Protection in China", *Managing the Chinese Environment* (Oxford: Oxford University Press, 1998), p. 33.

[11]Xiaoying Ma and Leonard Ortolano, *Environmental Regulation in China: Institutions, Enforcement, and Compliance* (Oxford: Rowman & Littlefield Press, 2000).

[12]*Ibid.*, p. 83.

[13]Vaclav Smil, *China's Past, China's Future: Energy, Food Environment* (New York: RoutledgeCurzon, 2004), p. 67.

developed a new model to estimate how much environmental air and water pollution costs China, including in terms of human health impact. It is the most comprehensive report on economic costs and human health impacts of environmental pollution undertaken in China. The report entitled *Cost of Pollution in China: Economic Estimates of Physical Damages* found out that the combined health and nonhealth cost of outdoor air and water pollution for China's economy came to 362 billion yuan in 2003 (or about 2.68% of the country's GDP).[14] According to this report, air pollution, especially in large cities, is leading to higher incidences of lung diseases, including cancer, respiratory system problems, and therefore higher levels of work and school absenteeism. The report concluded that regardless of income levels in China, the willingness to pay for reduced health risks associated with environmental pollution is about the same.

In its report entitled *Environmental Performance Review of China*, OECD (Organisation for Economic Co-operation and Development) gave a number of recommendations to China for environmental management, including establishing SEPA (State Environmental Protection Administration) as a ministry, making local leaders more accountable for their environmental performances and extending the use of pollution charges, user charges, emissions trading, and other market-based instruments.[15] In their paper published by *Science*, Jianguo Liu and Jared Diamond also suggested a reform of China's administrative system and a change in its model of economic development to tackle the environmental problems.[16] In the paper "The Great leap Backward? — The costs of China's Environmental Crisis",[17] Elizabeth C. Economy used a much harsher language to criticize China's mounting environmental problems, saying China's environmental

[14]The World Bank and SEPA, "Cost of Pollution in China: Economic Estimates of Physical Damages", The World Bank, Publications and Reports, March 2007, p. xvii, http://siteresources.worldbank.org/INTEAPREGTOPENVIRONMENT/Resources/China_Cost_of_Pollution.pdf.

[15]OECD Working Party on Environmental Performance, "Environmental Performance Review of China — Conclusions and Recommendations (Final), The OECD Publications and Documents, 9 November 2006, p. 5, http://www.oecd.org/dataoecd/58/23/37657409.pdf.

[16]Jianguo Liu and Jared Diamond, "Revolutionizing China's Environmental Protection", *Science*, Vol. 319, No. 5859 (2008), pp. 37–38.

[17]Elizabeth C. Economy, "The Great Leap Backward?", *Foreign Affairs*, Vol. 86, Issue 5 (2007), pp. 38–59.

problems "stem as mush from China's corrupt and undemocratic political system as from Beijing's continued focus on economic growth".[18]

Stefanie Beyer assessed the current state of China's environmental legislation framework in the paper "Environmental Law and Policy in the People's Republic of China", highlighting the surprising comprehensiveness of China's environmental protection legislation and enforcement tensions between the center and the periphery, a result of decentralization and growing local protectionism.[19] *The China Environment Yearbook (2005): Crisis and Breakthrough of China's Environment*,[20] known as the Green Book of Environment 2005 (*Huanjing Lupi Shu 2005 Nian: Zhongguo de Huanjing Weiji yu Tuwei*) in the original Chinese version, was written in English in 2007 as an attempt by "Friends of Nature", China's first environmental NGO, to gather a large number of experts to observe the current record and consider the process of environmental protection in China from the public perspective. *Environmental Governance in China*, by editors Neil Carter and Arthur P. J. Mol, published in 2007, gives special attention to the development of "environmental governance" in contemporary China, especially in the urban, industrial, and infrastructure sectors, showing how the rapid economic growth that has transformed China in recent years has major implications for the environment, as well as future economic development.[21] *Environmental Governance in China* was previously published as a special issue of the renowned journal of *Environmental Politics*, an internationally refereed journal sensitive to the distinction between goals of conservation and of a radical reordering of political and social preferences.

This book, with reference to the above-mentioned research materials, analyzes the factors in China's governance and political process that affect and restrain its capacity to handle the mounting environmental problems. It argues that solutions to China's ecological woes to a larger

---

[18] *Ibid*. p. 56.

[19] Stefanie Beyer, "Environmental Law and Policy in the People's Republic of China", *China Journal of International Law*, Vol. 5, No.1 (2006), p. 185.

[20] Liang Congjie and Yang Dongping (eds.), *The China Environment Yearbook 2005: Crisis and Breakthrough of China's Environment* (Boston: Brill, 2007).

[21] Neil T. Carter and Arthur P.J. Mol (eds.), *Environmental Governance in China* (London: Routledge, 2007).

extent lie in the political and institutional changes rather than more engineering, technological, and investment input. China is now at the crossroad of further political and economic reform, and the intensified public attention to environmental pollution may help the ruling Chinese Communist Party to decisively push forward the long-sluggish political reforms.

Chapter 1

Environmental Crises in China

CHINA ON THE BRINK OF WATER CRISIS

Severe Water Pollution

The rapidly deteriorating freshwater environment has become one of China's most urgent and vital ecological problems to cope with in the 21st century. Quite a number of books and articles have talked about this threat to China's sustainable development and national security. Among them, *China's Water Crisis* (*Zhongguo shui weiji*)[1] written by Ma Jun, a leading Chinese investigative journalist, was one of the earliest and most influential works touching this sensitive topic in China. In the *Zhongguo shui weiji* that was first published in Chinese in 1999 and later in English in 2004, Ma described in great detail the floods, water pollution, and scarcity problems existing in all seven of China's major drainage basins: the Yellow River, Yangtze River, Northwest, Northeast, North China Plain, Southeast, and Southwest. Through telling of the fresh information he collected himself after visiting many places in China, Ma told the readers that water pollution has become a nightmare that could haunt local people for years. He described a dire picture in the middle section of the Yangtze River, China's mother river: by the time the Han River (Yangtze's longest branch) reaches

[1]Ma Jun, *China's Water Crisis* (Norwalk: EastBridge, 2004).

the Wuhan City, its water is sometimes so filthy that it cannot even be handled by the local purification plant; one major purification plant on the Yangtze had to be moved 300 meters out into the rivers so that it could avoid taking in the water close to shore, which was untreatable because of pollutants coming from a large paper mill.[2]

In many parts of China, surface waters are severely polluted as a result of agricultural, industrial, and domestic discharges. Water quality is monitored by the State Environmental Protection Administration (SEPA) in about 500 river sections and by the Ministry of Water Resources in more than 2000 sections across the main rivers. China's lake and river water quality is classified from Grade I to Grade V based on concentrations of 30 substances, indicating good, fairly good, slightly polluted, poor, and hazardous. In 2004, about 25,000 kilometers of Chinese rivers failed to meet the water quality standards for aquatic life and about 90% of the sections of rivers around urban areas were seriously polluted[3] (Table 1.1). China's environmental monitors say that one-third of all river water, and vast sections of China's great lakes, the Taihu, Chaohu, and Dianchi, have water rated Grade V, the most degraded level, rendering it unfit for industrial or agricultural use.[4]

Many of the most polluted rivers and lakes have been void of fish for years. Among the 412 sections of the seven major rivers monitored in 2004, 42% met the Grades I–III surface water quality standard (that is, water that is safe for human consumption), 30% met Grade IV–V standards, and 28% failed to meet Grade V.[5] Water quality in the main river systems in northern China has been deteriorating rapidly in recent years. For all the five main river systems in the North (Songhua, Liao, Haihe, Huaihe and Huang rivers), sections with water ranked Grade IV and V — i.e.

---

[2] *Ibid.*, p. 82.

[3] The World Bank and SEPA, *Cost of Pollution in China: Economic Estimates of Physical Damages*, The World Bank, Publications and Reports, March 2007, pp. 4–6, http://siteresources.worldbank.org/INTEAPREGTOPENVIRONMENT/Resources/China_Cost_of_Pollution.pdf.

[4] Joseph Kahn and Jim Yardley, "As China Roars, Pollution Reaches Deadly Extremes," *The New York Times* (26, August 2007).

[5] The World Bank and SEPA, *Cost of Pollution in China: Economic Estimates of Physical Damages*, The World Bank, Publications and Reports, March 2007, p. 6, http://siteresources.worldbank.org/INTEAPREGTOPENVIRONMENT/Resources/China_Cost_of_Pollution.pdf.

Table 1.1. Supply of Surface Water that Does Not Meet Pollution Standards (million cubic meters).

Province	Urban Domestic Use	Rural Domestic Use	Industrial Use	Large Irrigation Use	Other Agricultural Use	Surface Total
Beijing	0	0	0	100	0	100
Tianjin	0	0	13	731	0	744
Hebei	0	0	14	910	1159	2083
Shanxi	30	69	105	342	612	1158
Inner Mongolia	13	0	67	468	194	742
Liaoning	41	0	381	2049	233	2703
Jilin	309	0	55	190	1084	1639
Heilongjiang	289	0	1190	354	2211	4045
Shanghai	1184	74	1018	0	646	2922
Jiangsu	707	609	2281	280	5629	9508
Zhejiang	379	215	1253	805	1011	3663
Anhui	133	254	134	0	1919	2440
Fujian	1	12	0	0	0	13
Jiangxi	4	0	141	0	67	211
Shandong	52	2	14	15	503	586
Henan	226	5	408	639	1396	2674
Hubei	101	5	334	10	54	504
Hunan	305	270	648	0	627	1850
Guangdong	1138	50	1276	0	100	2563
Guangxi	19	53	124	0	88	284
Hainan	0	0	0	0	0	0
Chongqing	69	69	103	0	36	277
Sichuan	247	126	150	0	73	596
Guizhou	1	0	50	0	1	52
Yunnan	0	0	0	0	0	0
Tibet	0	0	0	0	0	0
Shaanxi	4	3	0	290	0	297
Gansu	252	46	167	0	531	997
Qinghai	12	1	14	0	189	216
Ningxia	0	2	135	2162	1655	3953
Xinjiang	0	0	0	32	0	32
Total	5516	1866	10073	9348	20018	46821

Source: The World Bank and SEPA, *Cost of Pollution in China: Economic Estimates of Physical Damages*, The World Bank, Publications and Reports, March 2007, p. 88, http://siteresources.worldbank.org/INTEAPREGTOPENVIRONMENT/Resources/China_Cost_of_Pollution.pdf.

nonpotable water sources, but that may be used by industry (Grade IV) and agriculture (Grade V) — increased, while the better Grade I–III water — i.e. suitable for drinking water, swimming, and household use, and which also can support aquatic life — increased in the South.[6]

Pollution of seawater is also alarming. 30% of sites at which seawater quality is monitored have quality poorer than Grade III.[7] As large numbers of pollutants are infiltrating from land to the sea, the coastal marine ecosystem is worsening and the quality of ocean water is deteriorating. The most heavily polluted areas are concentrated along the coastline, and include Bohai Bay and the mouth of the Yangtze River. Lying off the coast of North China, one of the country's most populous and developed areas, the Bohai Sea has witnessed the highest increase of pollution levels. An area of 27,000 square kilometers accounting for 35% of Bohai's waters, failed to reach clean water standards.[8] Red tides occurred 96 times in 2004, and more than 20 were toxic.

In rural China, some 300 million people have no access to safe drinking water; the drinking water for 190 million people has harmful substances that exceed the prescribed limits; 63 million people drink high-fluoride water; 2 million people drink high-arsenic water; 38 million people drink high-salinity water, and some 11 million people in schistosome-stricken areas have no access to safe water.[9]

Pollution is also endangering China's groundwater, which provides 70% of the country's total drinking water. According to one report by the government-run Xinhua News Agency, the aquifers in 90% of Chinese cities are polluted.[10] This indicates the trend that water pollution is extending from surface to groundwater. Surface water- and air-pollutions are much easier to be detected than the contamination of groundwater, and long-term accumulation of contaminants will harm people's health

---

[6]*Ibid*.

[7]*Ibid*., p. 89.

[8]"Pollution Worsens in China's Sea Waters", *People's Daily Online*, 10, January 2005, http://english.peopledaily.com.cn/200501/10/eng20050110_170142.html.

[9]Liang Congjie and Yang Dongping (eds.), *The China Environment Yearbook 2005: Crisis and Breakthrough of China's Environment* (Boston: Brill, 2007), p. xxxv.

[10]Elizabeth C. Economy, "The Great Leap Backward?", *Foreign Affairs*, Vol. 86, Issue 5 (2007), p. 43.

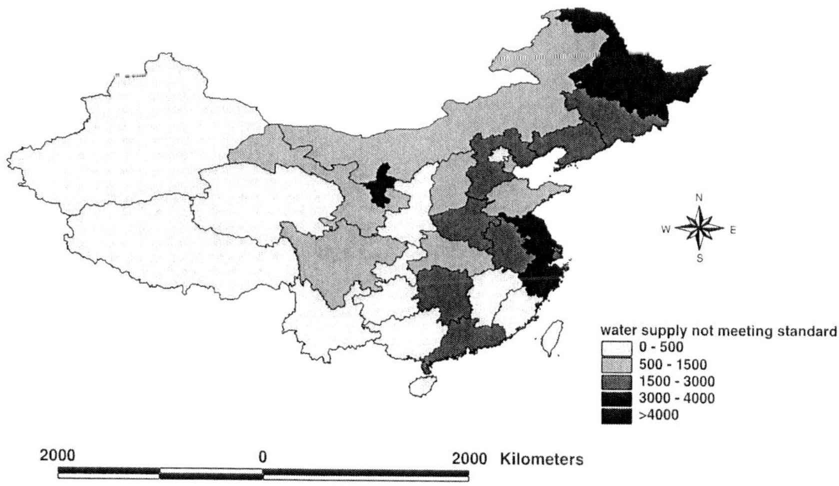

Figure 1.1. Polluted Water Supply in China (unit: million cubic meters).
Source: Ministry of Water Resource — The World Bank and SEPA, *Cost of Pollution in China: Economic Estimates of Physical Damages*, The World Bank Publications and Reports, March 2007, p. 89, http://siteresources.worldbank.org/INTEAPREGTOPENVIRONMENT/Resources/China_Cost_of_Pollution.pdf.

chronically. In China, nearly 700 million people drink water contaminated with animal and human waste.[11]

Figure 1.1 shows the volume of water supply (in millions of cubic meters) that does not meet the supply standards for each province. Polluted water in supply refers to water that exceeds the water quality standard relevant for the purpose.[12] The map was produced by the World Bank using the Ministry of Water Resources survey data from 2000 to 2003, and the multiyear coverage allows statisticians to account for annual variations due to rainfall and other factors.

In recent years, pollution-triggered water crisis broke out more frequently in some parts of China, arousing public panic and international attention. In November 2005, a blast at a chemical plant of

[11] *Ibid.*

[12] Polluted water for households refers to water worse than Grade III supplied (after treatment) to households; polluted water for industrial purposes refers to water worse than Grade IV supplied (after treatment) to industry; and polluted water for agricultural purposes refers to water worse than Grade V supplied to agriculture.

Jilin Petrochemical Company under the China National Petroleum Corporation (CNPC) sent an 80-kilometer-long slick of benzene into the Songhua River, which flows through Jilin and Heilongjiang provinces. The accident forced Harbin, the capital city of Heilongjiang, to suspend water supply from November 22 to 27 because of contamination. During that period, 29 industrial enterprises in the city stopped production and 23 reduced production, leading to a daily loss in output value of nearly 100 million yuan. Xie Zhenhua, then director of the SEPA, was pressured to resign after the crisis, becoming the highest-ranking official to step down for an environmental incident in China. In 2007 alone, blue-green algae outbreaks were reported in eastern Taihu Lake, Chaohu Lake, and southwestern Dianchi Lake, all threatening the local tap water supply and arousing nationwide attention.

The economic and social cost of water pollution is enormous in China. According to the World Bank, irrigation with polluted water costs 7 billion yuan per year, and the cost to fisheries was estimated at 4 billion yuan in 2003.[13] Two-thirds of the rural population is without piped water, which contributes to diarrhea disease and cancers of the digestive system. If a premature death is valued using a value of a statistical life of 1 million yuan, reflecting people's willingness to pay to avoid mortality risks, the cost of these health impacts is 1.9% of rural GDP.[14] There is no exact figure about the economic cost of the total water pollution, but many renowned research institutions including the World Bank estimated that it would at least worth 1% of China's GDP.[15]

China actually has a comprehensive legal framework for water resource and pollution management, with clear mechanisms to control abstractions and to set water quality objectives. The 2002 Water Law opens the way for integrated river basin management, stakeholder participation, and the use of market mechanisms in water management, in other words a major

[13]The World Bank and SEPA, *Cost of Pollution in China: Economic Estimates of Physical Damages*, The World Bank Publications and Reports, March 2007, p. xvii, http://siteresources.worldbank.org/INTEAPREGTOPENVIRONMENT/Resources/China_Cost_of_Pollution.pdf.

[14]*Ibid*, p. xiv.

[15]*Ibid.*, p. xvi.

reform of the water sector.[16] The deteriorating water pollution is mostly attributed to the rapid industrial development, loose enforcement of law, and regulations plus high-speed urbanization process in China. From 2000 to 2005, China's urban population increased by 103 million. This has likely contributed to increases in urban COD (chemical oxygen demand) and ammonia nitrogen loads.[17]

Water Shortage

China's water pollution and shortage are of highest concern among all environmental problems. Although China holds the fourth largest fresh-water resources in the world (after Brazil, Russia, and Canada), skyrocketing demand, overuse, inefficiencies, pollution, and unequal distribution have produced a situation in which two-thirds of China's approximately 660 cities have less water than they need and 110 of them suffer severe shortages.[18] According to Ma Jun, a leading Chinese water expert, several cities near Beijing and Tianjin, in the northeastern region of the country, could run out of water in five to seven years.[19]

Water shortage is a chronic problem that is closely related to pollution problems. China's water pollution is serious, extending from rivers, lakes to the coasts and spreading from surface water to groundwater, which exacerbates the shortage problem. At least 300 million Chinese lack adequate access to clean drinking water. Most of China's surface water is unfit for human consumption, and some of that not even clean enough for industrial use.

China's water shortage and pollution problems are more severe than any other large country in the world. China's total water resources have

---

[16]OECD Working Party on Environmental Performance, *Environmental Performance Review of China — Conclusions and Recommendations (Final)*, 9, November 2006, p. 5, http://www.oecd.org/dataoecd/58/23/37657409.pdf.

[17]The World Bank and SEPA, *Cost of Pollution in China: Economic Estimates of Physical Damages*, The World Bank, Publications and Reports, March 2007, p. 8, http://siteresources.worldbank.org/INTEAPREGTOPENVIRONMENT/Resources/China_Cost_of_Pollution.pdf.

[18]Elizabeth C. Economy, "The Great Leap Backward?", *Foreign Affairs*, Vol. 86, Issue 5 (2007), p. 42.

[19]*Ibid.*

decreased from 2785.5 billion cubic meters in 1997 to 2413 billion cubic meters in 2004. The 2004 per capita amount is approximately 2300 cubic meters, which puts China at 88th in the world.[20] Deserts cover 27% of the country's landmass, and the Yellow River runs dry so often and so long that some scientists have argued that it ought to be considered a seasonal phenomenon. By 2030, China will probably face a much more serious water situation, with per capita water resources quota to be only 1700 cubic meters, a generally acknowledged danger limit, when its population reaches 1.6 billion.[21]

AIR POLLUTION

Despite China's persistent effort to clean its air, air quality in many Chinese cities remains among the worst in the world. Chinese cities often seem wrapped in a toxic gray shroud, and ambient air pollution alone is blamed for hundreds of thousands of deaths each year.[22] Only 1% of the country's 560 million city dwellers breathe air considered safe by the European Union.[23] About 60% of cities above country level are likely to have met the grade II ambient air quality standard by 2005, and the SO_2 concentration in urban air, after dropping steadily since the early 1990s, began to climb again in 2002.[24] According to China's official statistics, its SO_2 emission rose from 19.9 million tons in year 2000 to 25.5 million tons in year 2005, jumping 28%.[25] The projected goal for 2005, however, was only 17.5 million tons

[20]Liang Congjie and Yang Dongping (eds.), *The China Environment Yearbook 2005: Crisis and Breakthrough of China's Environment* (Boston: Brill, 2007), p. 157.

[21]"China's Water Shortage to Hit Danger Limit in 2030", *People's Daily Online*, 16, November 2001, http://english.peopledaily.com.cn/200111/16/eng20011116_84668.shtml.

[22]Joseph Kahn and Jim Yardley, "As China Roars, Pollution Reaches Deadly Extremes", *The New York Times* (26, August 2007).

[23]*Ibid.*

[24]OECD Working Party on Environmental Performance, *Environmental Performance Review of China — Conclusions and Recommendations (Final)*, The OECD Publications and Documents, 9, November 2006, p. 5, http://www.oecd.org/dataoecd/58/23/37657409.pdf.

[25]China Environment Yearbook Editing Committee (ed), *China Environment Yearbook (2006)*, (Beijing: China Environmental Science Press, 2006), p.198 and China Environment Yearbook Editing Committee (ed), *China Environment Yearbook (2001)*, (Beijing: China Environmental Science Press, 2001), p. 125.

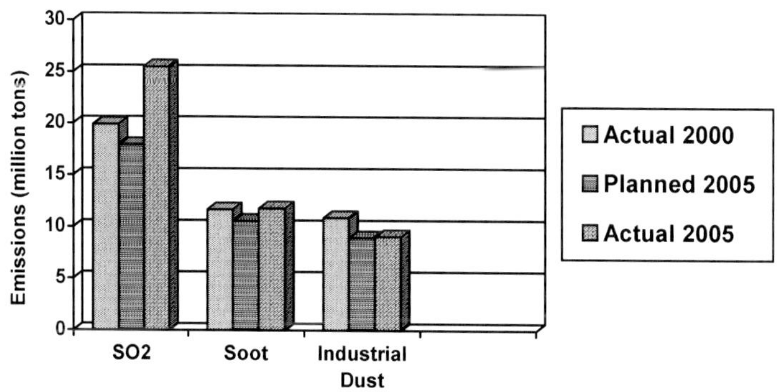

Figure 1.2. China's SO_2, Soot, and Industrial Dust Emissions (million tons).
Source: China Environment Yearbook Editing Committee (ed), *China Environment Yearbook (2006)*, (Beijing: China Environmental Science Press, 2006).

of SO_2 emission (Figure 1.2).[26] China is now the largest source of SO_2 emissions in the world. The World Bank estimated in its report that the economic burden of premature mortality associated with air pollution was 157.3 billion yuan in 2003, or 1.16% of GDP.[27]

Energy consumption, especially coal consumption, is the main source of air pollutants such as particles, SO_2, NO_x, and CO in most cities of China.[28] As the primary energy source, coal provides 69% of China's energy needs, while oil accounts for 21% and natural gas, hydro, nuclear, wind power, and solar energy altogether 10%.[29] The country is home to 16 of the world's 20 most polluted cities, and 4 of the worst off among them are in the coal-rich province of Shanxi, in northwestern China.[30] As much as 90% of China's SO_2 emissions and 50% of its particulate emissions are

---

[26] *Ibid.*

[27] The World Bank and SEPA, *Cost of Pollution in China: Economic Estimates of Physical Damages*, The World Bank Publications and Reports, March 2007, p. xiii, http://siteresources.worldbank.org/INTEAPREGTOPENVIRONMENT/Resources/China_Cost_of_Pollution.pdf.

[28] Ibid., p. 19.

[29] National Development and Reform Commission (NDRC), *China's National Climate Change Program*, China Climate Change Info-Net, June 2007, p. 9, http://www.ccchina.gov.cn/WebSite/CCChina/UpFile/File188.pdf.

[30] Elizabeth C. Economy, "The Great Leap Backward?", *Foreign Affairs*, Vol. 86, Issue 5 (2007), p. 40.

the result of coal use.[31] The rapid growth of China's transport industry especially the automobile sectors also contribute greatly to air pollution in the country. In 2006, China became the world's second largest car market and the third-largest car producer after the United States and Japan.

Although Beijing has spent over US$ 15 billion battling environmental pollution, more than it promised when the city won the Olympics bid in 2001, the capital's air quality remains poor and it has been an area of concern for the organizers, athletes, tourists as well as China's top leaders. Beijing is often blanketed by gray smog, some from coal-fired industries and construction sites, and others from neighboring provinces or rocketing vehicle numbers. Growing international concerns over Beijing's air pollution have embarrassed the organizers of the Games, imposing a huge challenge toward the "Green Olympics" concept put forward by Beijing years ago.

CLIMATE CHANGE

As the international community has become more preoccupied with the issue of climate change, China's rocketing increase in CO_2 emissions caught the spotlight of the outside world. As CO_2 is a normal component of the atmosphere and usually does no harm to human health, it is traditionally not regarded as a source of air pollution. The growing emissions of CO_2 and other greenhouse gases including methane and N_2O, however, have been making our planet warmer and warmer since the industrial revolution. This kind of climate change phenomenon mostly caused by human activities could lead to frequent meteorological disasters including drought, flood, and tropical storms, rising sea levels that may submerge coastal cities and island states and may even change the whole ecological system, nurturing the insect population and leading to epidemic outbreaks. To prevent such an ecological catastrophe, the international community has enacted and implemented the Kyoto Protocol, calling every nation to concretely cut their greenhouse gas emissions. The mitigation of climate change is a kind of public goods for the whole international community; so, those nations

[31] *Ibid.*

with largest carbon emissions in the world are definitely under close scrutiny and face huge moral pressures.

China is one of the nations under most focus during the long-lasting process of international climate negotiations. With CO_2-spewing, coal-fired power plants generating around 70% of China's seemingly exponential energy needs, the world's most populous nation is now firmly entrenched as one of Earth's most crucial climate changers. Anxiety about the gloomy climate prospect has made the public doubt about the country's GDP-first strategy and its industrial structure with focus on heavy and chemical industries. China has long been fending off an international accord that could force it to cut emissions and scale-back its economy. Beijing's fears of being targeted on climate change were heightened by the statement in April 2007 by an official of the International Energy Agency (IEA) that China could become the world's largest emitter of greenhouse gases as early as 2007. The Netherlands Environmental Assessment Agency (MNP) even announced that China's 2006 CO_2 emissions already surpassed those of the United States by 8%.[32] In 2006, global CO_2 emissions from fossil fuel use increased about 2.6%, which is less than the 3.3% increase in 2005. The 2.6% increase was mainly due to a 4.5% increase in global coal consumption, of which China contributed more than two-third.[33] With this, China topped the list of CO_2-emitting countries for the first time. In 2005, CO_2 emissions of China were still 2% below those of the United States.[34]

The global warming debate became more heated in 2007, at the 10th anniversary of the birth of the Kyoto Protocol that set mandatory targets for industrialized nations to cut emissions, as news reports released by the Intergovernmental Panel on Climate Change (IPCC) in 2007 were describing the connection between climate change and human activity with a high degree of certainty. The Kyoto Protocol, which is scheduled to expire by 2012, cannot meet the demand from the international community for further emission reductions. At the UN climate change summit

---

[32]Netherlands Environmental Assessment Agency, *China Now No. 1 in CO_2 Emissions; USA in Second Position*, Netherlands Environmental Assessment Agency Press Releases 2007, 19, June 2007, http://www.mnp.nl/en/dossiers/Climatechange/moreinfo/Chinanowno1inCO2emissionsUSAinsecondposition.html.

[33]*Ibid.*

[34]*Ibid.*

Table 1.2. China's Greenhouse Gas Emissions in 2004 (million tCO_2e).

Gas	World Total	China	% of World Total
CO_2	38,000	5050	13.3
CH_4	7000	720	10.3
N_2O	4000	330	8.3
Total	49,000	6100	12.5

Source: The statistics are respectively from National Development and Reform Commission (NDRC), *China's National Climate Change Program*, China Climate Change Info-Net, June 2007, p. 6, http://www.ccchina.gov.cn/WebSite/CCChina/UpFile/File188.pdf and Intergovernmental Panel on Climate Change (IPCC) Working Group III Report, *Mitigation of Climate Change (2007)*, p.103, http://www.ipcc.ch/ipccreports/ar4-wg3.htm.
Note: PFC, HFC, SF_6 figures are omitted.

in Bali of Indonesia in December 2007, new round of international negotiations formally started to discuss a new accord substituting the Kyoto Protocol. China, whose greenhouse gas emissions have been rising rapidly together with its double-digit economic growth, is now facing more international pressures for stringent emission reduction targets (Table 1.2). If China cannot handle this problem properly in the future talks, its soft power may be weakened with international images damaged. Without any detailed emission-cutting obligations in the Kyoto Protocol, China's GHG emissions have been growing very rapidly due to its heavy reliance on coal and booming of its nascent auto industry. As a carbon-intensive fuel, coal contains almost twice the amount of carbon per unit of energy compared to natural gas, and about 20% more than petroleum. In fact, there are more coal power plants installed in China than in the United States, the United Kingdom, and India combined. In Beijing, the capital, the auto number doubled in just five years to reach 3 million, with more than 1000 new private cars hitting on the roads everyday.[35] In the capital city, transport

[35]Bill Schiller, "*Beijing, the Triathlon of Commutes*", *The Star* (10, December 2007), http://www.thestar.com/News/World/article/284051.

already accounts for 20% of society's total energy consumption, and frequent traffic congestion exacerbates the problem, making these vehicles less efficient with more emissions when queuing on the streets.

China itself is facing a potentially devastating impact from climate change, which has become a social and economic problem in the country. China admitted in its first national program on climate change that "annual average air temperature has increased by 0.5°C to 0.8°C during the past 100 years, which was slightly larger than the average global temperature rise".[36] The prolonged droughts will lead to agricultural output reduction, while the acceleration of glacier melting on the northwestern Qinghai–Tibet plateau will reduce water supply to Yangtze and Yellow Rivers, China's two longest rivers. Besides, China's sea level rising triggered by global warming is faster than the global average, and this will impose threat to China's coastal cities, the nation's economic powerhouse.

OTHER PROBLEMS

China has made remarkable progress in the reuse of waste materials and safe disposal in landfills. China implemented a cleaner production law in 2003 and revised its 1995 waste law in 2004, building up a complicated legal framework for modern waste management. The concepts of the "3Rs" (reduce, reuse, recycle) and of the "circular economy" (*xunhuan jingji*) have become parts of China's 11th Five-Year Plan (2006–2010).

Nevertheless, the amount of municipal waste, industrial waste, and hazardous waste far exceed what can safely be treated and disposed of.[37] Most of this solid and liquid hazardous waste is not effectively regulated, and it is polluting soils and contaminating groundwater aquifers and rivers. Some of the waste that is stored and awaiting treatment continues to pile up at dump sites. Many of these sites are unlined and thus unable to prevent

[36]National Development and Reform Commission (NDRC), *China's National Climate Change Program*, China Climate Change Info-Net, June 2007, p. 4, http://www.ccchina.gov.cn/WebSite/CCChina/UpFile/File188.pdf.

[37]OECD Working Party on Environmental Performance, *Environmental Performance Review of China — Conclusions and Recommendations (Final)*, The OECD Publications and Documents, 9, November 2006, p. 7, http://www.oecd.org/dataoecd/58/23/37657409.pdf.

contaminants from spreading to the surrounding environment, particularly to nearby aquifers.[38]

Several factors contribute to difficulties in controlling contamination from dumps and landfills: information on the location and condition of solid waste disposal facilities is scarce; many factories store solid waste on site, where they often escape control by environmental authorities; and, the expert knowledge needed to measure and control hazardous waste is limited.[39] The 10th Five-Year Plan target of increasing the capacity of municipal landfills to 150 kt/day was not achieved, and human health and the environment are put at risk through a proliferation of uncontrolled dumps surrounding the cities.[40]

China has established an adequate legal framework for protecting the ecological system and biodiversity. The country actively reports on its international commitments and also publishes annual state of the environment reports related to its internal goals and targets. Protected areas at the national, provincial, prefecture, and county levels have been dramatically increased, and China has received international recognition for its wetlands, biosphere reserves, and natural and cultural heritage preservation programs.[41]

There is still much to do, however, to further develop shelter forests, protect biodiversity, and to promote farm forestry on land sensitive to soil erosion. Various environmental protection programs with China have begun to recognize the value of environmental outreach including alien species and endangered wildlife, but there is insufficient monitoring to assess trends and evaluate the protection status of natural reserves.[42] Although China has a relatively high percentage of total area classified as protected, marine habitats and species are not sufficiently represented and are subject to land-based sources of pollution and habitat alteration,

---

[38]Xiaoying Ma and Leonard Ortolano, *Environmental Regulation in China: Institutions, Enforcement, and Compliance* (2000), p. 4.
[39]Eduard B. Vermeer, "Industrial Pollution in China and Remedial Politics", *The China Quarterly*, Vol. 156 (1998), p. 981.
[40]OECD Working Party on Environmental Performance, *Environmental Performance Review of China — Conclusions and Recommendations (Final)*, The OECD Publications and Documents, 9, November 2006, p. 7, http://www.oecd.org/dataoecd/58/23/37657409.pdf
[41]*Ibid.*, p. 8.
[42]*Ibid.*

in addition to exploitation pressures. Little has been done to promote biodiversity protection on forestland and to tailor payments to forest owners to the provision of forest ecosystem services, and China has not yet ratified the Bonn Convention on migratory species, although it is active in regional cooperation on migratory waterbirds.[43]

[43] *Ibid.*

Chapter

2

Bureaucratic Structures on Environmental Protection

The long-time fruitless effort in curbing Taihu Lake's pollution is only one example showing China's inefficiency in solving local ecological problems. Among various causes that may cover a wide range of realms, the most fundamental one may lie in the multilayered and multisectioned governmental structures in the country. In a state-centric country like China, the government rather than the market plays a dominant role in environmental protection; thus, when the governmental institutions have innate flaws, the environmental problem will probably become incurable.

INTERACTION AMONG THE PARTY, THE ADMINISTRATION, AND THE LEGISLATURE

China's unique one-party political system predetermines the country's governance quality on many certain issues. Concerning the interrelations among the ruling party, the legislature (congress), and the cabinet (state council), the Chinese Communist Party (CCP or the Party) always plays a predominant role in all decision-making processes with its members penetrating into the legislature and the cabinet and its ruling status stipulated by the Constitution. It is efficient to enforce central policies in various parts of the vast territory at various levels under such unitary political system, but this also causes intractable governance problems such as lack of supervision, corruption, and loose law enforcement. In terms of environmental governance, the growing attention paid by the Party at the central level to the pollution and other ecological problems has mobilized the whole governmental system to make more effort in the green area, but the monolithic power structure from top down to the grass root in the multilayer structure prevents such problem from being effectively solved when local party bosses are not incentivized to push forward the green causes.

Although the three systems (the Party, the administration, and the legislature) at various levels are all led by party members, there is still an implicit pecking order of their power in daily political life. Party secretaries of CCP committees at all five governmental levels (central, provincial, municipal, county, township) have predominant say in the affairs within the realm, while the administration chiefs at various layers are responsible for routine and detailed administrative work but have to listen to the party committee's voices during decision-making on important or controversial issues. The system of People's Congress (legislature) at all levels is the least powerful branch among the three. The National People's Congress (NPC), tabbed by western media as "robber stamp", is regarded by the Constitution as the nation's top legislative body, enacting both basic environmental law and special environmental laws. In practice, however, the State Council (administration) drafts proposed laws and refers them to the NPC and its Standing Committee. The State Council plays this influential legislative role because it sits at the top of a vast bureaucracy that includes virtually all ministries, commissions, and other administrative agencies, and thus it has considerable access to expertise and resources.[1] People's Congress at various levels usually holds plenary sessions once a year, and delegates have very limited power in supervising the law enforcement in practice. China lacks an independent judicial system, with courts at all levels under the absolute leadership of the corresponding party committees.

[1]Xiaoying Ma and Leonard Ortolano, *Environmental Regulation in China: Institutions, Enforcement, and Compliance* (Oxford: Rowman & Littlefield Publishers, 2000), p. 14.

Under such systems, the party committee and the administration are in the upper hand while the legislature and courts are a bit marginalized in the game of power and thus lack the authority for check and balance. Theoretically speaking, the party secretary, administration chief, and chairman of the people's congress are at the same hierarchical level and have the same rank in the officialdom, but in practice, the party secretary is more powerful than the administration chief and often concurrently acts as the chairman of the people's congress. The courts are always ranked lower than the above three systems at all levels, and a special political and jurisprudence subcommittee under the party committee have the leadership over local court, procuratorate, and police. Without sufficient supervision, selective law enforcement can become rampant and *guanxi*, or social connections, is essential for polluters to elude punishment from the administrative department such as the environmental protection bureaus or the courts. If entrepreneurs keep good relations (*guanxi*) with local officials through bribery in forms of cash, valuable gifts, invitation to banquets, and all-expense-paid trips abroad, it is not difficult for them to avoid compliance of strict environmental standards and thus lower the manufacturing cost. The key figures for polluting companies to maintain good *guanxi* with are those cadres in the local party committee, the administrative departments, and the courts. For the environmental administrative fines or pollutant discharge fees collected by the local environmental protection bureaus, it is possible for the related entrepreneurs to bargain with the administrative staff due to lack of supervision.

CENTRAL-LEVEL ADMINISTRATIVE SYSTEM ON ENVIRONMENTAL PROTECTION

China's governmental structure is highly hierarchical. Each government unit above the village level has an administrative (or bureaucratic rank), which is an important reflection of power and status.[2] The Chinese government is made up of a number of functional units (ministries, commissions, administration, bureaus, and departments), which are in charge

[2]Ibid., p. 34.

of various issues such as foreign affairs, trade, education, public health, finance, land resources, economic projects, personnel, and environmental protection. A ministry or a commission has the same administrative rank (*buji*) as a provincial-level government. These functional units exist at the vertical chain through successively lower territorial levels of government. Individual functional units within this system receive administrative guidance from their parent units above them; they are also subject to the leadership of the local governments to which they belong; but communication between functional units at the same territorial level has traditionally been very limited.[3]

At the central level, before the establishment of the Ministry of Environmental Protection in March 2008, the State Environmental Protection Administration (SEPA) had been taking charge of the nation's environmental protection issues for about 10 years, with its sub-branches established at provincial, municipal, county, and even township level under the names of Environmental Protection Bureaus (EPBs) or Offices (EPOs) (Figure 2.1). Unlike the powerful Ministry of Foreign Affairs or the National Development and Reform Commission (NDRC), the ministerial-level SEPA, despite enjoying the same rank, was actually called an "Organization Directly Under the State Council", (*guowuyuan zhishu jigou*) rather than a real-sense ministry or commission under the State Council. The predecessor of the SEPA used to be part of the Ministry of Construction in the 1980s and was promoted to the ministerial level in 1998 as environmental problems received increasing attention in China.

The SEPA received most of its funding from the State Council, and was supported in its efforts to formulate national policy by the Committee on Natural Resources and Environmental Protection under China's top legislature, National People's Congress. The SEPA has proved to be a weak unit inside the State Council, compared with other ministries or commissions in full sense, such as the NDRC, the Ministry of Commerce, or the Ministry of Agriculture. China's "development first" strategy makes those economic departments inside the government play a dominant role in the

[3] Abigail R. Jahiel, "The Organization of Environmental Protection in China", in Edmonds, Richard Louis (ed), *Managing the Chinese Environment* (Oxford: Oxford University Press, 1998), p. 34.

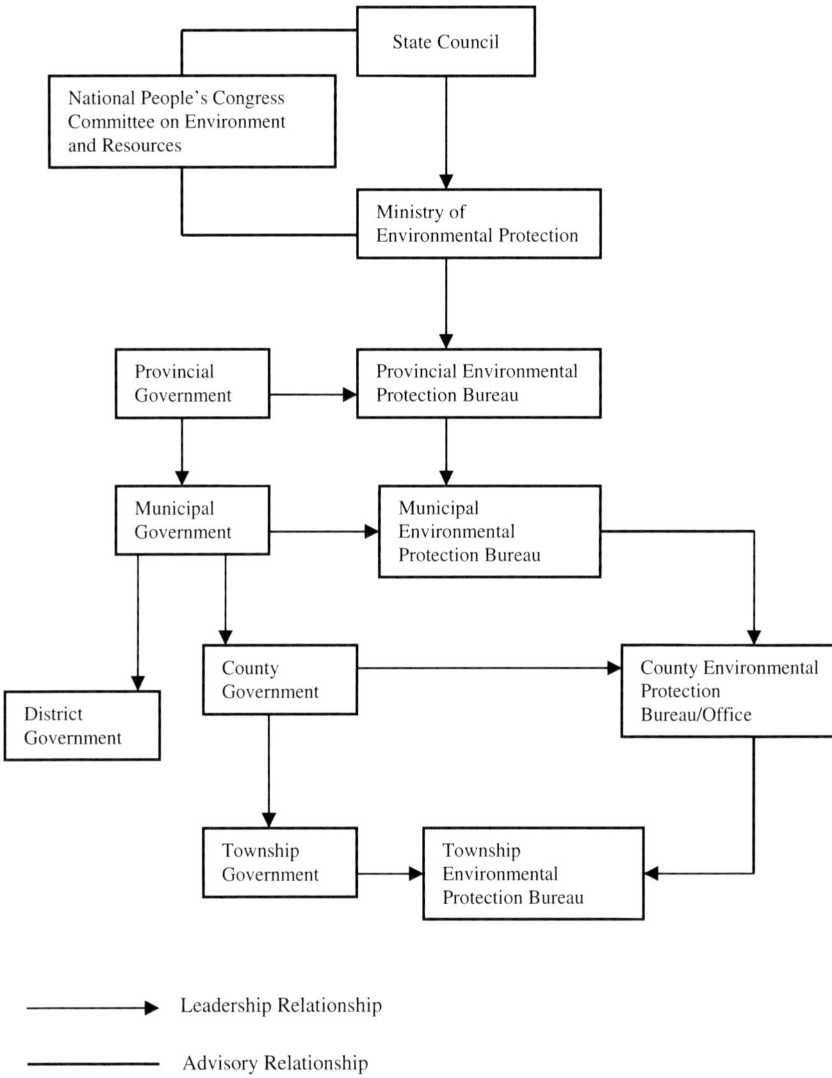

Figure 2.1. China's Environmental Protection Apparatus.

inter-agency decision-making games while those like the SEPA and China Meteorological Administration (CMA) play minor roles.

As SEPA was not a full-fledged ministry, its representatives usually could not join cabinet-level decision-making process through the two normal forms of State Council conferences: the State Council Executive Conference and the State Council Plenary Conference. Every week, the State Council holds its regular executive conference to discuss current issues, with the participation of the premier, vice premiers, state councillors, and the secretary-general to the State Council. The State Council Plenary Conference will also include chiefs of ministries, commissions, the central bank, and the National Audit Office inside the cabinet. Hence, the inclusion of a governmental department in the cabinet is crucial for its influences upon the decision-making at the central level.

Moreover, pollution control functions in China used to be scattered among several ministries including water resources, construction, agriculture, oceanic, and forestry administration. It is the Ministry of Foreign Affairs that is in charge of talks on international environmental treaties, and it is the National Development and Reform Commission (NDRC) that is responsible for the environmental protection industry, energy efficiency, climate change, and carbon emission cutting issues. The protection of water resources including rivers and lakes is the business of the Ministry of Water Resources, while the Bureau of Forestry takes care of ecological protection in the woods and forestry. The environmental protection of oceans, rural soil and water, and urban water and garbage treatment is respectively in the hands of the Oceanic Administration, the Ministry of Agriculture, and the Ministry of Construction. The Ministry of Finance is in charge of governmental fund investment in environmental protection, while the Meteorological Administration is authorized to carry out nationwide air quality supervision. Although three ministries, namely the Ministry of Water Resources, the Ministry of Construction, and the Ministry of Agriculture have been involved in urban and rural water management, it is the Ministry of Health that takes charge of the drinking water safety issues in cities and countryside. The SEPA, therefore, actually had very limited powers when dealing with pollution control and other environmental issues.

In 1998, China established the National Coordination Committee on Climate Change, an inter-agency organization that now includes representatives from the NDRC, Foreign Ministry, Ministry of Science and Technology, CMA, SEPA, and other 10 ministerial-level departments. With NDRC Director Ma Kai appointed Chairman of the National

Coordination Committee on Climate Change, the NDRC takes charge of all routine affairs in the Committee and obviously has the overpowering role in the decision-making process. Compared with the SEPA and CMA that are potentially sympathetic to the more pro-active approach of emission abatement, the NDRC leans toward rapid economic growth as well as materialized benefits including fund and technology transfers.

Inter-agency conflicts between the NDRC and SEPA have become obvious as domestic environmental problems are exacerbated. The latest evidence of such inconsistency is the disarray in China's car fuel policy. Vigorously pushed by the SEPA, the central government had planned, on 1, July 2007, to introduce a standard for cars known as Euro III, which reduces carbon monoxide and other emissions. On 1, July 2007, however, the NDRC said the rules should be delayed for up to 2 years, because it said the lack of compatible fuel could damage new engines.

The heavy-weighted role of Foreign Ministry and Ministry of Science and Technology enhances NDRC's advantageous position in the climate change decision-making process. The Foreign Ministry puts an emphasis on the preservation of national sovereignty, meaning that developing countries have the need and the right to develop, while the Ministry of Science and Technology focuses on foreign technological aid. Without any change in such inter-agency mechanism on environmental protection, the "development-first" views of those advantageous economic sectors will continue to affect China's effort to cope with ecological deterioration.

WEAKNESS IN LOCAL ENVIRONMENTAL PROTECTION DEPARTMENTS

Besides the central level, environmental protection bureaus at various local levels are also in a disadvantageous position when performing their duties. In all cases, it is the local government, not the higher levels of the environmental protection apparatus, that provides local environmental agencies with their annual budgetary funds, approves institutional advancements in rank, and determines increases in personnel and even allocation of resources such as cars, office buildings, and employee housing.[4] Local

[4]Ibid., p. 35.

leaders, including the party bosses and administrative chiefs, all put economic growth in a top priority position and are unwilling to see local economy and investment hampered by implementation of strict environmental regulations. High-speed economic growth not only is an important index to measure local officials' performances linked with their future job promotions, but also increases local fiscal revenues and benefits local officialdom materially as a whole. China's maintenance of double-digit economic growth, which is much higher than expected by the government in its 11th five-year program, is imposing huge pressure upon the natural resources and ecological environment.

Since early 1990s, China's officialdom has formed a culture whereby the GDP growth is a major indicator to measure administrative chiefs' performances and even determines their future job promotions. In 2004, the Chinese government started to work on the criteria and indexes of a "Green GDP", which deducts the cost of environmental damage and resource consumption from the traditional gross domestic product. In 2007, however, China announced that it would scrap "Green GDP" report, saying "there is no internationally accepted standard for calculating green GDP."[5] The initiative has proved too sensitive to continue after the three-year trial period, because it met strong resistance from China's local officials, who have been long time rewarded largely on their ability to improve growth.

As one of the most affluent provinces in China, Jiangsu does not lack sufficient fund or technology for environmental protection. The Taihu algae crisis, however, showed that its environmental governance was no better than other parts of China. Officials at all levels have been aware of the seriousness of Taihu's problem for a long time, but local administration's unitary goal on GDP growth greatly weakened clean-up motivations. It is EPB/EPO's job to supervise environmental causes, but their heavy dependence upon local government and disadvantaged positions in interaction with other departments made green targets very difficult to achieve.

Without a unified and universally-reinforced environmental standard, the pollution can only be transferred, instead of being fundamentally solved. Statistics show if the national standard is strictly followed, a chemical project with annual sales of 5 million yuan needs to invest about

[5]Shai Oster, "China Scraps Green GDP Report", *Wall Street Journal* (17, July 2007).

1 million yuan in infrastructure construction for wastewater treatment, often exceeding the initial investment over the project.[6] Facing huge pressures from the central government and the public, local officials had to take high-profile measures including closing down pollution factories and punishing lower-level officials, but their anxiety about the possible slowdown of economic growth and huge cleanup cost is still there. After some time, when the public attention gets shifted, the environmental standards and law enforcement will become loose again. This is the vicious circle that keeps going on and prevents Taihu Lake from turning limpid again. Local stricter environmental requirement could only drive away entrepreneurs and investors if other regions of China remain the same.

Many local EPBs have for years been facing the problem of fund shortage, and some even cannot afford the salaries of their staff. Local environmental agencies therefore retained a portion of pollution fees and fines they collected from polluters to alleviate the hunger for fund. As these arrangements might foster corruption and encourage EPBs to focus on collection of pollution fees and fines instead of reducing total pollution, the country decided to reform this revenue retention system through introducing the scheme called "two separate lines for revenues and expenses" (*shouzhi liangtiao xian*). During this nationwide reform that started in late 1990s, not only environmental departments, but also all kinds of administrative agencies at all levels were required to decouple their revenue from the expense streams in a bid to prevent corruption and promote accountability. In environmental management, all pollution fees and fines now go directly to the local finance bureau, and the EPB is no longer guaranteed a partial refund of their pollution fees.[7] Taken the Guangzhou City as an example, under the new system, all fees and fines collected become part

[6]"Out of the Plight: Taihu Lake After the Wuxi Water Pollution", (*zouchu weiju: wuxi shuiwuran shijianhou de taihu*), *People's Daily Online* (15, August 2007), http://paper.people.com.cn/dd/html/2007-08/15/content_19740890.htm.

[7]Carlos Wing-Hung Lo and Shui-Yan Tang, "Institutional Reform, Economic Changes, and Local Environmental Management in China: the Case of Guangdong Province", in Neil T. Carter and Arthur P. J. Mol (eds.), *Environmental Governance in China* (London: Routledge, 2007), p. 48.

of the general funds, with 10% going to the national treasury, 5% to the provincial treasury, and the remaining 85% to the municipal treasury.[8]

The new system requires EPBs to develop an annual budget, which must be approved by the finance bureaus and then incorporated into the regional budget before submission for parliamentary approval. Different from past practices in which each bureau collected its own fees and used the funds at its discretion, the new system to a large extent has played a positive role in curbing corruption and increasing transparency. For some environmental officials, however, the new system also has a negative impact on the EPB because financial power is now concentrated at the finance bureau, and it takes more time and effort for the EPB to request funding.[9] From this perspective, the new system has made EPBs more financially dependent upon the local government.

The new system did not solve the funding insufficiency problem effectively in all parts of China. In less developed regions with limited financial budget, the funds allocated to local EPBs each year sometimes are far from sufficient to cover the expenses of normal operation. The difference had to be offset by local EPBs themselves through providing some forms of commercial services and retaining a portion of such income while the majority of it is submitted to financial bureaus. In practice, many EPBs still keep about one-third of the pollution fee to make its ends meet although it conflicts with the principle of "two separate lines for revenues and expenses". How the revenue-expense-separation system has been implemented therefore varies in different localities.

EPB's fund shortage is also partly attributed to the overstaffing problem in local governmental departments. In the EPB of Lingbi County, Anhui Province, the number of formal staff members on the state payroll (*bianzhinei renyuan*) was 21 in 2007, but in reality, this EPB recruited a total of 94 staff members, with only 8 purely reliant upon county payment and 13 relying 70% of their salaries on county payment.[10] All the other 73 persons (*bianzhiwai renyuan*) were working for the

---

[8] *Ibid.*

[9] *Ibid.*, p. 49.

[10] "Lingbi: Overstaffing in the EPB and Pollution Fee Not Enough for Making a Living", *People's Daily Online* (31, July 2007), http://nc.people.com.cn/GB/6049455.html.

environmental supervision team (*huanjing jiancha dadui*), with their income utterly dependent upon the EPB instead of the county government. The Lingbi EPB in 2006 collected pollutant discharge fee worth 990,000 yuan, 35% of which was retained by the EPB to pay the salary for its staff.[11] Despite such practice that actually conflicted with the State Council Regulation on Pollutant Discharge Fee Collection and Use, the average income of staff working in the environmental supervision team was still only a little higher than 200 yuan per month, even lower than the 360-yuan-per-month minimum salary level set by the Lingbi government.[12] The overstaffing problem is partially due to the rampant nepotism in local government department, with major county officials putting their relatives and friends into the EPBs and other government agencies. Such severe funding shortage prevents local EPBs from making all-out efforts to decrease total pollutant emissions and instead forces them to focus on collecting pollution fees and fines. With the revenue and expense gradually being separated under the new system, some of these pollution fees and fines may turn into bribes of various forms. If the power and functions of environmental agencies at various levels are expanded as suggested by some scholars and experts, corruption involving environmental officials may increase if relevant supervision and monitor system cannot follow.

CLASH BETWEEN GDP GROWTH AND POLLUTION CONTROL

Since China started its reform and opening-up process in 1978, the government has been regarding the economic growth as its priority task. Over the past 30 years, the ruling communist party kept using the high-speed gross-domestic product (GDP) growth and increase of people's living standard to legitimize its one-party governance in the country. Since 1992, when Deng Xiaoping, general-designer of China's reform and open-up policy, advocated the introduction of market economy during his tour to southern China, the officialdom in the country started to form the GDP-first culture, in which the GDP growth rate became a key index to evaluate local

[11] *Ibid.*
[12] *Ibid.*

officials' performances and even decide their promotions. From 1978 to 2006, China's GDP growth kept an amazingly-high growth of 9.7% per year, jumping from 364.5 billion yuan to 21.09 trillion yuan. In 2005, China's GDP had already exceeded that of France and Britain, ranking fourth in the world, only after the United States, Japan, and Germany.

The high-speed economic growth together with the accelerated industrialization and urbanization process in China has imposed huge resource and environmental pressure upon the country's feeble ecological systems. Actually China's pollutant discharge does not always follow the trend of economic growth. Table 2.1 shows that industrial waste water and gas discharged in the country dropped remarkably from 1990 to 1996 when the nominal per capita GDP jumped more than 200%. Only after 1996 did the major discharged pollutants begin to rebound again and grow simultaneously with the GDP.

After China's accession to the World Trade Organization (WTO), its export to other parts of the world kept an amazingly high growth rate

Table 2.1. China's Industrial Pollutants and GDP Figures (1990–2005).

Year	Industrial Wastewater Discharge (billion tons)	Industrial Waste Gas Emission (million tons)	Industrial Solid Waste (billion tons)	Per Capita GDP (yuan)	GDP (billion yuan)
1990	24.9	36	0.58	1644	1866.8
1991	23.57	35.15	0.59	1893	2178.2
1992	23.39	33.72	0.62	2311	2692.4
1993	21.95	33.11	0.62	2998	3533.4
1994	21.55	27.31	0.62	4044	4819.8
1995	22.25	28.71	0.65	5046	6079.4
1996	20.59	27.17	0.47	5846	7117.7
1997	22.67	49.22	1.06	6420	7897.3
1998	20.05	37.07	0.8	6796	8440.2
1999	19.73	35.89	0.78	7159	8967.7
2000	19.42	26.58	0.82	7858	9921.5
2001	20.26	23.98	0.89	8622	10965.5
2002	20.72	23.07	0.95	9398	12033.3
2003	21.23	38.51	1	10542	13582.3
2004	22.11	38.91	1.2	12336	15987.8
2005	24.31	40.35	1.34	14040	18308.5

Source: China's National Bureau of Statistics.

which turned itself into the world's workshop for all kinds of labor-intensive products. Since 2003, China's GDP growth started to accelerate to more than 10% annually, with its output in iron and steel, cement, aluminum, and other heavy and chemical industry soaring and fixed asset investment becoming overheated. After Wen Jiabao became premier in 2003, he began to worry about this economic overheating that put huge pressure upon the country's natural environment and brought about power and raw material shortages. Despite a series of the so-called "macro-control" policy measures since 2004, the overheating problem could not be effectively alleviated while the bottleneck of natural resources plus the environmental pollution became even more serious. Table 2.1 shows that from 1990 to 2001, before China joined the WTO, its per capita GDP more than quadrupled from 1644 yuan to 8622 yuan, while its industrial wastewater discharge reduced from 24.9 billion tons to 20.26 billion tons, down 19%, and industrial waste gas emission dropped from 36 million tons to 23.98 million tons, decreasing by 33%. Only the industrial solid waste rose from 0.58 billion tons to 0.89 billion tons during this period, increasing by 50%.

After 2001, however, all the three pollution indexes picked up momentum together with the speedup of GDP growth. From 2001 to 2005, China's per capita GDP jumped 63% from 8622 yuan to 14,040 yuan, while its industrial wastewater discharge, industrial waste gas emission, and industrial solid waste rose by 20%, 68% and 51%, respectively to 24.31 billion tons, 40.35 million tons, and 1.34 billion tons. These figures have shown that pollution discharged from China's industrial sectors increased dramatically after China's entry into the WTO in 2001, when the country's economy was strongly boosted by its export with a more favorable external trade environment. China's economy has been restructured after it became more and more dependent upon exportation. The title of the world's workshop not only means that the country has been one of the largest beneficiaries from the economic globalization process and trade liberalization, but also implies that it might be the victim of the side effects of such a globalization trend as well. One of the most worrisome side effects from the globalization and related neo-liberalism theory is the environmental price that the whole human society has to pay for the development of such a global market economy.

Trade facilitation and market economy on a global scale follow the logic of "winner takes all" and push every participant to lower production cost by all means. When enjoying cheaper goods, consumers tend to increase their demand and consume more than before, which surely leads to more consumption of natural resources during the manufacture process and more pollution to the ecological system. Facing fierce global competitions, some countries will lower their domestic standards for environmental and labor protection to allure more foreign fund and lower the production costs for export commodities. Unfortunately, China, a country with flawed governance systems and impatient eagerness to reap benefits from the accelerated globalization process, had to swallow the torment of ecological degrade and environmental pollution simultaneously when billions of dollars of revenue flew into the country.

ENVIRONMENTAL IMPACT OF CHINA'S WTO MEMBERSHIP

The above analysis shows that 2001, which is also when China formally joined the WTO, became a watershed for the country's pollution control. The discharge volume of major pollutants and carbon dioxide emissions[13] changed their previous flat momentum and started to jump after the country's WTO accession, a landmark movement to further liberalize China's foreign trade, accelerate economic growth, speed up opening-up of many domestic business sectors, and increase the country's international prestige.

For a long time, divided views existed over the environmental implications of the WTO membership and trade liberalization. On one side are those who view free trade primarily as a catalyst for environmental improvement and argue that trade liberalization[14]:

1. it leads to economic growth, which is a precondition for acquiring the technical and institutional capability to address pollution (*the wealth argument*);

---

[13]For details of China's CO_2 emissions, please refer to the figure "China's CO_2 Emissions from 1980–2005" in Chap. 10 — "How China Manages and Benefits from CDM Projects".

[14]Abigail R. Jahiel, "China, the WTO, and Implications for the Environment", in Neil T. Carter and Arthur P.J. Mol (eds.), *Environmental Governance in China*, (London: Routledge 2007), p. 163.

2. it reduces costs of imports and thus provides greater access to cleaner production techniques (*the technology argument*);
3. it removes subsidies and trade barriers, encouraging structural changes to the economy and exchange on the basis of comparative advantage and efficient use of resources (*the structural change/efficiency argument*); and
4. it promotes environmental legal reform to prevent rejection of products for sale overseas (*the "regulatory race to the top" argument*).

On the other side are those who view trade liberalization as antithetical to environmental protection and argue that[15]:

1. the expansion in output comes at a cost to the environment (*the scale argument*);
2. the comparative advantage can negatively affect the environment if a country's industrial structure and natural endowments favor specialization in pollution-intensive industries (*the "pollution haven" argument*); and
3. trade liberalization encourages the reduction of environmental standards as countries compete for foreign investment (*the "regulatory race to the bottom" argument*).

During China's insistent effort to join the General Agreement on Tariffs and Trade (GATT) and its successor, WTO, experts and scholars had conducted plentiful of research on the implications of WTO membership for the economy, society, politics, and ecological environment. Realizing that the economic globalization process had met strong civic resistance in many parts of the world for reasons of labor and environmental protection, the Chinese government tried to gain more domestic support for this sensitive membership application by asking the academia to embrace neoclassical economics and depict a rosy picture for the post-WTO era. The negative ecological impact of the WTO membership, therefore, was ignored amidst optimistic projections that industrial pollution would decrease after the accession.

China's accession to the WTO means reduction of both its import tariffs and other members' tariffs upon its exportation, plus end of export

[15] *Ibid.*, p. 164.

subsidies, import quotas, and open of certain sectors to foreign investment. It is true that the WTO membership has brought China with unexpectedly high economic growth, accelerating inflow of high technology and improvement of its legal system, which enhanced the developing nation's capacity to address environmental problems. What really matters for the environment, however, are the total scale of the economy and the direction in which the economy is heading for. Although China has made more or less progress in energy and resource efficiency with stricter environmental standards since the WTO accession, the rapidly expanding output in quantity offset such improvement and caused substantial growth of pollution in total amount. The boom in China's manufacturing sector, the largest beneficiary from tariff reduction, has contributed a large part to the increase of air and water pollution and soil degradation. Rocketing export strained domestic supply of raw materials and energy, salvaging many environmentally destructive ventures (such as small-sized coal mines) that should be phased out under strict environmental regulations.

China's textile industry, which has significant ecological impact due to heavy chemical pollution and water consumption, witnessed its export to other parts of the world soar after 2001. China's textile export grew 16%, 28% and 21%, respectively in 2002, 2003, and 2004,[16] thanks to reduction of other countries' tariffs and Chinese tariffs for imported dyes, cotton, and textile machinery. Like the coal industry, within the textile industry small firms have prospered, accounting for about 85% of all cotton textile manufacturers.[17] The Chinese government has found it difficult to regulate and monitor these small polluting enterprises that are often equipped with inferior technology, due to a shortage of rural environmental protection personnel and to economic pressures to approve projects quickly and overlook environmental factors.[18] To compete in the international market and continuously cut production cost, Chinese textile producers have tried to reduce salary cost and pollution control expense all the time.

[16]"Integration of Textile Trade and China's Textile Export", (*fangzhiping maoyi yitihua yu zhongguo de fangzhiping chukou*), Zhangjiagang Economic and Trade Commission (30, May 2005), http://www.zjgjm.gov.cn/ArticleShow.asp?ArticleID=2010.
[17]*Ibid.*, p. 170.
[18]*Ibid.*

With sewer discharge per unit of product, almost double the amount discharged in developed nations, China's textile sector is a major source of the country's industrial sewage.

WTO membership also facilitates China to seek cheap raw materials in other countries for processing and manufacturing at home, which produces spillover ecological impacts. Since WTO-entry, China has been able to meet its rapidly rising domestic demand for wood products and has become a major international base for furniture production and wood processing.[19] The reduction of timber tariffs and elimination of quotas have helped Chinese merchants to fell wood in voracious quantity and buy at very low prices in countries with poor forest management like Myanmar, Cambodia, Indonesia, and Russia. Besides timber, Chinese companies are also busy seeking other raw materials and fossil fuels in places like the African continent, West Asia, Southeast Asia, and South America. By investing in various mining and oil exploration ventures there, certain environmental conditions in China may improve as the country's economy continues to grow (as happened with forests), but the voracious appetite of the Chinese manufacturing economy and its overseas customers from the industrialized world serves to exacerbate environmental problems elsewhere, much as development within the North has for decades brought ecological destruction to countries within the South and to the global ecosystem more broadly.[20]

[19] *Ibid*, p. 175.
[20] *Ibid*.

Chapter 3

Environmental Legal Systems in China

BASIC ENVIRONMENTAL LAW

The People's Republic of China (PRC)'s environmental legal system did not start to develop until the country officially adopted reform and opening-up policy in 1978. Hardly any progress was achieved in environmental law-making between the establishment of PRC in 1949 and the end of "Cultural Revolution" in 1976. China, for the first time, enacted an environmental protection provision into its basic national law in 1978, when the State's responsibility for the protection of the environment was added to the Constitution.[1] Articles 9 and 26 of China's 1982 Constitution call upon the state to ensure the rational use of natural resources, protects rare animal and plants, protect and improve the living environment and the ecological environment, and eliminate pollution and public hazards. In 1979, a tentative version of China's basic environmental law, the PRC Environmental Protection Law for Trial Implementation, was promulgated by the National People's Congress Standing Committee. China's top legislature removed the trial implementation status of the basic environmental law by enacting the PRC Environmental Protection Law in 1989. Before

[1] Stefanie Beyer, "Environmental Law and Policy in the People's Republic of China", *China Journal of International Law*, Vol. 5, No. 1 (2006), p. 192.

1989, many Chinese authorities believed that only the National Environmental Protection Agency and environmental protection bureaus were responsible for preventing pollution. With the passage of the 1989 law, governments at the national, provincial, city, and county levels were given explicit environmental protection responsibilities.[2]

China's basic environmental law contains the following four guiding principles[3]:

- *Coordination of environmental protection and economic development.* National environmental protection plans must be integrated into national economic development plans; in addition, the state should adopt economic and technical measures that are environmentally beneficial (Article 4).
- *Pollution prevention.* While measures to abate existing waste discharges are being undertaken, future releases should be prevented, mainly by using enhanced environmental management at new pollution sources (Articles 13, 15, and 26).
- *Polluter responsibility.* Waste dischargers should bear the costs of cleaning up environmental problems they cause; thus polluters should pay discharge fees on waste releases that exceed discharge standards, and they should invest in waste treatment facilities (Articles 24, 27–29, and 31).
- *Strengthening environmental management.* Because reducing pollution by investing in new waste treatment facilities is expensive, the first line of attack is to cut waste releases by improving management procedures (Articles 9 through 15).

The Environmental Protection Law established a system for environmental management, monitoring, liability, and enforcement to address water, air, solid waste and noise pollution. These include general requirements for discharge registration systems, for the levying of fees, for environmental impact assessments as well as measures for the control and elimination of pollution supplemented by provisions for legal

---

[2]Xiaoying Ma and Leonard Ortolano, *Environmental Regulation in China: Institutions, Enforcement, and Compliance* (Oxford: Rowman & Littlefield Publishers, 2000), p. 16.
[3]*Ibid.*

liability.[4] The Environmental Protection Law stipulates the basic principles and guidelines concerning environmental protection in China.

SPECIALIZED ENVIRONMENTAL LAWS AND REGULATIONS

Besides the basic environmental protection law, China has enacted many special environmental laws that cover almost all the fields related to environmental protection (Table 3.1). The most notable laws include the Law on the Environmental Impact Assessment (2002), Law on Conserving Energy (1997), Law on Prevention and Control of Pollution from Environmental Noise (1996), Law on the Prevention and Control of Environmental Pollution by Solid Waste (1995), Law on the Prevention and Control of Atmospheric Pollution (1995), Law on the Protection of Wildlife (1988), Water Law (1988), and Law on Prevention and Control of Water Pollution (1984).

China's chief executive organ — the State Council — and its ministries, commissions, and departments such as the State Environmental Protection Administration (SEPA) have issued numerous administrative edicts to implement environmental policies stipulated in the basic and special environmental laws.[5] Some examples of these administrative edicts include the Regulations on the Administration of Medical Wastes (2003), the Implementation of the Law on the Water Pollution Prevention and Control (2000), the Interim Regulations on the Prevention of Water Pollution in the Huaihe River Valley (1995), and the Regulations on the Control Over Dumping Wastes into the Sea Waters (1985). Local legislature, the people's congresses, and their respective standing committees have the right to enact local environmental regulations (*difang fagui*), while local governments are entitled to issue local administrative rules (*difang xingzheng guizhang*) implementing national environmental laws and regulations.

[4] Stefanie Beyer, "Environmental Law and Policy in the People's Republic of China", *China Journal of International Law*, Vol. 5, No. 1 (2006), p. 193.

[5] The State Council enacts administrative regulations (*xingzheng fagui*) while its ministries, commissions, and departments issue administrative rules (*xingzheng guizhang*).

Table 3.1. Major Special Environmental Laws in China

Special Environmental Law	Year of Issuance
Law on Radioactive Pollution Prevention and Control	2003
Law on the Environmental Impact Assessment	2002
Law on the Promotion of Clean Production	2002
Law on the Administration of Sea Areas	2001
Law on Desert Prevention and Transformation	2001
Law on Conserving Energy	1997
Law on Prevention and Control of Pollution from Environmental Noise	1996
Law on the Prevention and Control of Environmental Pollution by Solid Waste	1995
Law on the Prevention and Control of Atmospheric Pollution	1995
Law on Water and Soil Conservation	1991
Law on the Protection of Wildlife	1988
Water Law	1988
Land Administration Law	1986
Mineral Resources Law	1986
Fisheries Law	1986
Forestry Law	1985
Grassland Law	1985
Law on Prevention and Control of Water Pollution	1984
Marine Environment Protection Law	1982

ENVIRONMENTAL DISPUTE SETTLEMENT

With the sizzling economic growth in China over the past three decades, people have been witnessing more and more environmental disputes among the government, individuals, and corporations. From the mid-1980s to 1997, the number of environmental disputes in China had been maintained at about 100,000 per year, but after 1997, such numbers started to rise dramatically, reaching about 600,000 in 2005. The quality of a country's environmental protection largely depends on how effective the state can protect its citizens' legal rights and interests from being infringed by other parties' pollution activities. Environmental cases involve civil, criminal, and administration litigation. In a legal and political milieu that traditionally refrains from the encouragement of individual rights, many environmental grievances are not transformed into disputes, but rather endured.[6]

Political stability and social harmony are always the priority concerns of both Chinese administrative and judicial departments; so, their focus is often on pacifying the disputes through conciliatory approaches rather than safeguarding individuals' legal rights and interests, especially when individual rights are sometimes considered to disturb the social order. In some specific environmental cases such as noise disputes, relevant parties are advised to resort to the approach of people's mediation (*renmin tiaojie*). The participation in mediation is voluntary; however, settlement agreements are, according to the newly issued provisions on people's mediation, treated as civil contracts, implying that the agreement may be neither modified nor repudiated.[7] As most environmental disputes are regarded as administrative matters, the administrative authorities instead of the courts play a decisive role in the solution process. In reality, environmental protection bureaus at various levels take charge of most environmental disputes. The administrative authorities are empowered to make a final decision, including the imposition of liability for compensations when mediation fails, and if one party refuses to accept the decision, it may bring a suit before the people's court.[8] In practice, however, parties involved in pollution cases in China seldom resort to litigation approaches mainly due to the counts' lack of independence from administrative departments and party committees. Local courts usually are expected to have the similar ruling as the final decision made by local administrative authorities, and the lawsuit cost may be too high for plaintiffs to bear. In fact, although the number of environmental disputes has been growing rapidly in recent years, that of environmental lawsuits in China did not increase at a corresponding rate.

Moreover, some state acts that may have environmental impact are exempted from being sued when they are deemed by the government to be vital for national interest and security. These cases include the controversial Three Gorges Dam Project and North–South Water Project that may have potential negative impact upon the ecological system.

---

[6] Stefanie Beyer, "Environmental Law and Policy in the People's Republic of China", *China Journal of International Law*, Vol. 5, No. 1 (2006), p. 190.

[7] *Ibid.*

[8] *Ibid*, p. 191.

It is not a usual case that an environmental dispute will evolve into criminal prosecution, which may result in imprisonment or other severe punishment for the accused. If the final ruling by both the administration and the courts cannot satisfy the victims from pollution cases, they can still depend on China's special mechanism of "letters and visits" (*xinfang zhidu*) and go to the *xinfang* office at the upper level to file complaints about local miscarriages of justice. This system provides the last opportunity for disadvantaged people to express their grievances about pollution, but as the *xinfang* office is established inside the government, its priority concern is still social stability rather than defending individual rights. Its role in handling the disputes fairly is therefore questionable and limited.

STATUTORY AND ENFORCEMENT FLAWS

China actually has one of the world's most complicated environmental law systems, with basic environmental law as the backbone and other national laws in various special areas as important complementary parts. Besides, the State Council, China's chief executive organ, and its ministries and departments, are entitled to enact administrative regulations and rules on specific environmental issues. The Ministry of Environmental Protection, National Development and Reform Commission, the Ministry of Water Resources, the Ministry of Agriculture and State Forestry Administration are all authorized to issue environmental regulations. Local courts and governmental departments across China also have the right to publicize detailed regulations for implementing the above-mentioned environmental laws and regulations at the central level according to special local conditions. All these central and local laws, regulations, and rules constitute China's environmental legal system in a broad sense.

Although China has so many laws and regulations, most of them were written in a rush and hence used quite vague and imprecise language. Some laws and regulations are more like policy statements than real laws. Typically, actions are encouraged but rarely required and even where concrete duties are stated, only little guidance is provided on procedures and specific goals.[9] According to Article 13 of the Law on Prevention and Control

---

[9] *Ibid*, p. 205.

of Water Pollution, "the environmental impact statement (on water pollution) shall include views of units and residents where the construction project is to be located,"[10] but no further details are provided if the nearby residents' opinions are not considered. According to Article 14 of the same law, "the pollutant discharging units mentioned in the preceding paragraph shall report in time if any substantial change occurs in the categories, quantities or concentrations of the water pollutants discharged."[11] There is also no clear definition about the "substantial change" and no corresponding punishment if the polluters do not follow the regulation.

Despite the large number of special laws and regulations concerning environmental protection, China's environmental legal system still misses some important aspects. Up to the present, no policy has been adopted that requires polluters to restore degraded property and to recover its pre-degradation status.[12] Neither have clean-up standards for contaminated soil been issued nor has liability in relation to toxic waste remediation been introduced.[13]

When designing the environmental legal regime, Chinese lawmakers expected local governments to play a very cooperative role in enforcing those laws and regulations. In practice, however, this high degree of administrative cohesion does not exist, and the laws fail to anticipate the possibility that certain government interests might diverge sharply from those of the environment department and create a major obstacle to strict enforcement of both national and local environmental legislation.[14] As stakeholders of local polluting companies, local authorities in many cases

---

[10] "Law of the People's Republic of China on Prevention and Control of Water Pollution, Article 13", Ministry of Environmental Protection of The People's Republic of China (11, May 1984), http://english.mep.gov.cn/Policies_Regulations/laws/environmental_laws/200710/t20071009_109915.htm.

[11] "Law of the People's Republic of China on Prevention and Control of Water Pollution, Article 14", Ministry of Environmental Protection of The People's Republic of China (11, May 1984), http://english.mep.gov.cn/Policies_Regulations/laws/environmental_laws/200710/t20071009_109915.htm.

[12] Stefanie Beyer, "Environmental Law and Policy in the People's Republic of China", *China Journal of International Law* Vol. 5, No. 1 (2006), p. 207.

[13] *Ibid.*

[14] *Ibid.*

tend to help these companies ignore environmental standards and regulations that may hinder investment inflow and economic growth. Fees and fines are rarely determined authoritatively; instead, they are often negotiated and fall far below the cost of damage that the harmful activity has caused, as well as below expenses for pollution control facilities.[15] For a long time, the maximum fine for the illegal pollutant discharge activities by enterprises in China was maintained at the level of 100,000 yuan, much lower than the cost of installing cleanup facilities in many cases. Even this standard for fines is leniently implemented in practice and many county-level enterprises can easily find leeway through bribing local officials.

[15] *Ibid.*

Chapter

4

Environmental NGOs and Emerging Civil Society

The establishment of Friends of Nature in 1994 kicked off the rapid growth of environmental nongovernmental organizations (ENGOs) in China. Such organizations, a symbol of China's emerging civil society, have become a major source of environmental knowledge as well as champions of public participation in environmental protection. Although ENGOs have become increasingly visible players in China's green politics, their roles and functions are still quite restraint under China's current state-centric political systems, and therefore, they are not as strong and influential as their international peers. Past experiences in western countries since 1960s show that significant supervision and participation from NGOs, the media, and other civil society groups are requisites for better environmental governance; so, if China wants to address the ecological problem more efficiently, it has to improve the institutional environment for the vibrant growth of ENGOs and effective media surveillance. Meanwhile, in interaction with the traditional mass media, the Internet, and international NGOs, indigenous ENGOs can serve both as sites and agents of democratic social change in China.[1]

[1] Yang Guobin, "Environmental NGOs and Institutional Dynamics in China", *The China Quarterly*, No. 181 (2005), p. 46.

THE DEVELOPMENT OF ENVIRONMENTAL NGOs IN CHINA

When the Chinese government started to deepen its reform and opening-up policy in the 1990s, civil activities became frequent and all kinds of NGOs working for public and special interests mushroomed. With increasing awareness about environmental protection, people set up ENGOs engaged in environmental education, nature conservation, species protection, policy advocacy, and many other activities. Friends of Nature, China's first ENGO, was established in 1994 when only nine ENGOs were founded in China, including four student organizations. By 1996, the number had grown to 28, including 10 student organizations. The number increased dramatically from 1997 to 1999, when at least 69 ENGOs were founded, 43 of which were student organizations.[2] China's official statistics showed that by the end of 2005, there were a total of 2768 ENGOs in the country, with 224,000 employees, 69,000 of whom were full-time and 155,000 part-time.[3] A total of 2.977 billion yuan (US$ 425 million) were raised by these 2768 NGOs in 2005,[4] and the numbers of ENGOs and their employees are expected by the All-China Environment Federation, a national ENGO with strong government backgrounds, to grow by 10%–15% annually in China.[5]

Generally speaking, China's environmental NGOs now can be categorized into seven types: registered NGOs, nonprofit enterprises, unregistered voluntary groups, web-based groups, student environmental associations, university research centers, and government-organized NGOs (GONGOs) (Table 4.1). In the state-centered country, GONGOs definitely enjoy much more financial advantages and policy convenience compared to other NGOs, but this kind of government background also restrains its role as a third force against the state and market.

The large number of national and local environmental GONGOs, to some extent, represents the government's endeavor and willingness to fight against pollution. Due to administrative restraint and focus on

---

[2] *Ibid*, p. 50.

[3] "China now has 2768 Environmental NGOs", *Xinhua News* (23, April 2006).

[4] *Ibid*.

[5] "Rocky Road to a Brighter Environment", *China Daily* (3, July 2006).

Table 4.1. Main Types of Environmental NGOs in China.

Organizational type	Registration status	Examples
Registered NGOs	Registered as social organizations (*shehui tuanti*) or private, nonprofit work *units* (*minban fei qiye danwei*)	Friends of Nature; Green River
Nonprofit enterprises	Registered as business enterprises but operate as nonprofit organizations	Global Village of Beijing; Institute of Environment and Development
Unregistered voluntary groups	Unregistered organizations that function as NGOs	Green Earth Volunteers
Web-based groups	Unregistered groups that operate mainly through the Internet	Green-Web; Greener Beijing
Student environmental associations	Registered with campus Youth Leagues yet function and perceived as NGOs	Sichuan University Environmental Volunteer Association
University research centres/institutes	Affiliated with institutions of higher learning but operate as NGOs	Centre for Legal Assistance to Pollution Victims; China University of Political Science and Law
Government-organized NGOs (GONGOs)	Social organizations established by government agencies, also known as state-owned NGOs (SONGOs)	China Environmental Science Association

Source: Yang Guobin, "Environmental NGOs and Institutional Dynamics in China", *The China Quarterly*, No. 181 (2005), p. 50.

academic activities, environmental GONGOs lack passion and incentives to push forward the environmental protection on certain sensitive issues. Compared to other environmental NGOs in China, however, some of these GONGOs such as the China Association of the Environmental Protection Industry, the China Society of Environmental Sciences, the China Energy Conservation Association, and the Chinese Society for Sustainable Development are more professional and boast expertise in environmental planning, decision-making, enforcement ability, and project application. Some of the incumbent or retired officials become heads of these GONGOs, and their efficient communication channels with other government departments facilitate the routine activities in promoting environmental protection causes. Pan Yue, Vice Minister of the State Environmental Protection Administration, predecessor of the Ministry of Environmental Protection, acted as the leader of the China Environmental Culture Promotion Association (CECPA) for a long time. Pan, a famous outspoken official for his critics on China's environmental management system, used his influence in the officialdom to make the CECPA play an active and sometimes aggressive role in the semi-civil activities, which produced positive social impacts. The main activities of this association include training universities' environmental protection associations to enhance capacity building, holding forums on environmental protection, and promoting communications and interactions with grassroots organizations. In April 2005, CECPA developed an environmental training program in association with the Global Village of Beijing and the State Environmental Protection Administration.[6]

Other kinds of ENGOs in China, with a common aim to promote environmental consciousness and public participation, avoid confrontational methods and adopt approaches that encourage learning, cooperation, and participation. Normal activities include public lectures, symposiums, field trips, newsletter announcement, on-line discussions, and even Internet petition. Although some organizations are edging their way into more contentious areas, the general tendency has been to shy away from radical issues and tactics.[7]

While avoiding confrontational tactics, ENGOs in China often adopt the boundary-spanning strategy, which takes advantage of political gray zones to advance contestable claims and extract concessions and often "employs the rhetoric and commitments of the powerful to curb political or economic power."[8] Under China's authoritarian regime, ENGOs

---

[6] Fu Tao, "Development of Environmental NGOs in China", in Liang Congjie and Yang Dongping (eds.), *The China Environment Yearbook 2005: Crisis and Breakthrough of China's Environment* (Boston: Brill, 2007), p. 300.

[7] Yang Guobin, "Environmental NGOs and Institutional Dynamics in China", *The China Quarterly*, No. 181 (2005), p. 52.

[8] Kevin J. O'Brien, "Neither Transgressive Nor Contained: Boundary-Spanning Contention in China", *Mobilization*, Vol. 8, No. 1 (2003), p. 53.

sometimes have to use the government's own words such as sustainable development and environmental protection as a weapon of protest, resistance, and collective actions. Legal action is a common practice by ENGOs to protect pollution victims and fight polluting industries.

China's environmental NGOs started their missions from environmental enlightenment education, and later more and more became involved in concrete environmental protection, green community development, legal assistance, and exchanges with other NGOs. Friends of Nature was the first ENGO to introduce the mobile van for environmental education in China. Their van, named Antelope, traveled across urban and rural areas to popularize participatory environmental education in primary and middle schools.[9] In 1999, Global Village of Beijing took a garbage classification and sorting initiative to develop the concept of a green community.

After 2000, more and more ENGOs started to use the Internet as a useful weapon to bypass the government's media censorship and impose pressure upon some sensitive ecological issues. In 2003, more than 10 ENGOs launched a joint advocacy effort to ask the government to reconsider possible negative ecological effect of the South-to-North Water Diversion Project. Due to the serious power shortage in the summer of 2004, Global Village of Beijing and other five domestic and international ENGOs introduced a joint on-line initiative in Beijing, calling on the air-conditioned temperatures in hotels, restaurants, and offices to be kept at 26°C.

GREENING OF MEDIA AND INTERNET

The Chinese media have been strong supporters of ENGOs from the very beginning, partly due to the "greening" of the media.[10] Many media professionals directly participate in ENGOs as their organizers or members. Green Camp, Green Earth Volunteers, Green Plateau, Tibetan Antelope

[9]Fu Tao, "Development of Environmental NGOs in China", in Liang Congjie and Yang Dongping (eds.), *The China Environment Yearbook 2005: Crisis and Breakthrough of China's Environment* (Boston: Brill, 2007), p. 301.

[10]Wen Bo, "Greening the Chinese Media", *China Environment Series*, Issue 2 (1998) pp. 39–44.

Information Center, Tianjin Friends of Green, and Panjin Black-Beaked Gull Protection Association are all led by journalists or former journalists, while Friends of Nature has some influential journalists in its membership.[11] Due to the insufficient support from the government and market resources, some NGOs regard the media as powerful amplifiers of their voices. As environmental issues are often controversial and easy to attract eyeballs, the media like to put these events in the headlines, and this is exactly what the ENGOs want. In interaction with these ENGOs, some journalists get more involved in the environmental movement and become sympathetic and supportive of the green causes. Internationally speaking, the alliance between the media and ENGOs is quite common, but in China, with special political system, journalists and reporters are in an awkward position, sandwiched between the Party superiors who have political authority over the press and the market, which puts economic constraints on it.[12] Thus, the media share structurally similar positions with ENGOs, and such structural homology fosters a reciprocal relationship between the media and the ENGOs.[13]

The media coverage and public controversies surrounding dam-building on the Nu River and its potentially dire ecological consequences showed how effective the media could be in helping ENGOs to alter the government's agenda. According to an incomplete statistics, from August 2003 to September 2004, this controversial issue became the focus of more than a hundred articles and programs by many influential newspapers, magazines, radio, and television outlets. The message in these reports grew increasingly critical, as they changed from positive comments to dissenting opinions, as revelations about the negative impacts of the dam surfaced.[14] Diverse reporting perspectives and heated debates in the mass media channeled citizen opinions to influence government policies. Some professional reporters renounced their role as independent and

[11] Yang Guobin, "Environmental NGOs and Institutional Dynamics in China", *The China Quarterly*, No. 181 (2005), p. 56.

[12] Hong Liu, "Profit or ideology? The Chinese Press between Party and Market", *Media, Culture and Society*, No. 20 (1998), p. 37.

[13] Yang Guobin, "Environmental NGOs and Institutional Dynamics in China", *The China Quarterly*, No. 181 (2005), p. 56.

[14] Fu Tao, "Development of Environmental NGOs in China", in Liang Congjie and Yang Dongping (eds.), *The China Environment Yearbook 2005: Crisis and Breakthrough of China's Environment* (Boston: Brill, 2007), p. 295.

objective observers and became sympathizers for disadvantaged people and marginalized NGOs.[15] Finally, in April 2004, China's State Council halted the hydropower project being planned on the Nu River in Yunnan Province. China's Premier Wen Jiabao, reportedly cited "a high level of social concern" as an important reason for suspending the dam-building project. Such a reversal after public criticism has hardly been typical of the Chinese government — nor was the nature of the public criticism typical.[16]

Compared with traditional media, the Internet has the advantages of speed, broad reach, and interactivity. The number of Chinese Internet users has been growing rapidly since the year 2000. In the first half of 2002, the number of netizens in China reached 45.8 million, jumping more than 30% from half a year ago. By the end of 2007, about 210 million Chinese people became Internet users, very close to the number of 215 million netizens in the United States. The year 2008 has witnessed China boast the world's largest number of Internet users, with a large number of rural people having access to the Internet. Despite political controls and censorship from the government, Chinese Internet users still find it more and more convenient to resort to Internet for public expression and political activism, especially against the backdrop of strict restriction over traditional media.

Now most ENGOs in China have their own web sites. Internet provides the platform to share information and knowledge, and some ENGOs maintain active mailing lists that send environmental information regularly to subscribers by e-mail. Internet is being used by individual ENGOs to expand their capacity and engage in dialogues with other interest groups. When the voices from some marginalized organizations are amplified through Internet, they tend to attract more social attention and have a larger influence upon policy-making process. Presently, information exchanges between NGOs through e-forums and e-newsletters are becoming increasingly frequent, and joint advocacy targeting focal issues demonstrates the intensive interaction among NGOs.[17] Many ENGO web

---

[15] *Ibid.*

[16] Yang Guobin and Craig Calhoun, "Media, Civil Society, and the Rise of a Green Public Sphere in China", *China Information*, Vol. 21 (2007), p. 211.

[17] Fu Tao, "Development of Environmental NGOs in China", in Liang Congjie and Yang Dongping (eds.), *The China Environment Yearbook 2005: Crisis and Breakthrough of China's Environment* (Boston: Brill, 2007), p. 296.

sites also have bulletin boards for public discussion, and blogs have become a new online form for individuals to express their views, release news, and publicize pictures and videos concerning environmental issues. Together with the rapid growth of ENGO web sites in number and influence, more and more environmental activists are creating their personal homepages on the Internet. The Internet's networked nature facilitates and accelerates the information spread among numerous web sites, including both the related environmental web sites and nonrelevant ones. When just one or two ENGO web sites or personal homepages start to discuss certain environmental issues, these issues are likely to spread quickly to other web sites and become the limelight of traditional media like newspapers and television.

Besides Internet, other modern telecommunication means such as short-messages (SMS) through mobile phones can also provide nontraditional channels for information circulation on sensitive environmental issues. With increasing public awareness about environmental issues, sometimes local people will voluntarily join collective actions to safeguard their legitimate rights that might be threatened by possible pollutions. In May and June 2007, several hundred people staged a peaceful demonstration against a planned chemical plant named Tenglong Aromatic PX Company in southern China's Xiamen City. The work was suspended by the government after residents sent about 1 million mobile phone SMS in protest against possible health dangers.[18]

POLITICAL CONSTRAINTS AND CHALLENGES FACING CHINESE ENGOs

Policy Constraints

The Chinese government has to some extent encourage the development of NGOs for handling increasing social problems. Chinese high-level officials have praised ENGOs such as Friends of Nature and called for support for the healthy development of these organizations. The government

[18]"Protests Against Xiamen Chemical Plant Continue Despite Decision to Suspend", *AP News* (1, June 2007).

promulgated two regulations in 1998 to provide the regulative framework for the registration and management of NGOs and nonprofit organizations. These regulations have both positive and negative impact upon the development of NGOs in China.

On the one hand, they represent official recognition of the legal status of such organizations. On the other hand, they also contain very restraint articles. According to the regulations, any social organization to be established in China needs to have a sponsoring governmental institution before being registered to the Ministry of Civil Affairs. This presents a major hurdle to registration because an NGO is considered a liability not an asset to its sponsoring institution.[19] Besides, the regulations stipulate that in the same administrative area, there should not be more than one organization for any specific area of work (such as environmental protection). In many cases, GONGOs have already filled that fixed quota.[20] Even Liu Qi, a member of the Chinese Communist Party Politburo and Party Secretary of Beijing, complained in March 2008 that such compulsory linkage between the government and social organizations has hampered the normal development of social organizations.[21] Figures show that the number of NGOs per 10,000 citizens is 110 in France, 97 in Japan, and only 2.1 in China.[22] The total expense for China's NGOs only accounts for less than 1% of its gross domestic product (GDP), much lower than the world's average 4.6% and developed countries' average of 7%.[23]

Moreover, due to the government's strict control and restrictions over NGOs, it is difficult for NGOs to recruit enough capable personnel. It is also not easy for most NGOs to obtain funding from private or state-owned

---

[19] Jin Jiaman, "The Growing Importance of Public Participation in China's Environmental Movement", in Jennifer Turner and Fengshi Wu (eds.), *Green NGO and Environmental Journalist Forum: A Meeting of Environmentalists in Mainland China, Hong Kong, and Taiwan* (Washington, DC: The Woodrow Wilson Center, 2001), pp. 5–8.

[20] Yang Guobin, "Environmental NGOs and Institutional Dynamics in China", *The China Quarterly*, No. 181 (2005), p. 55.

[21] "It's difficult to find mother-in-law: 20-year obstacle for NGOs", *Southern Metropolitan News* (*Nanfang Dushi Bao*) (16, March 2008).

[22] Yang Dongpin, "China's Environmental Protection at the Crossroad," (*shizilukou de zhongguo huanjing baohu*), China Internet Information Center (22, March 2006), http://www.china.com.cn/chinese/zhuanti/hjwj/1161880.htm.

[23] *Ibid.*

companies in the country. Many ENGOs have to resort to foreign foundations, international ENGOs, and multinational corporations. This in many occasions increases the government's distrust over some ENGOs, suspecting that such funding from overseas sources may have political intentions of subverting the government. As most ENGOs in China cannot raise fund in the way like their foreign peers, many face the problem of financial shortage that hamper their normal operation and routine activities.

Weak Capacity Building and Internal Management

Although most Chinese ENGOs have become more mature and sophisticated after more than a decade of development, they still face insufficiency and weakness in their managerial structure and professional capacity, which prohibits them from playing a larger role in boosting the social environmental protection causes. Despite the growing public environmental concern in China, the actual environmental knowledge of Chinese citizens is still quite low, which restrains the further development of ENGOs.

Many of China's first generation of grassroots ENGOs are spontaneous in management and function, and depend too much on their founder-leader. Leaders' personal ambitions and charisma shaped these organizations and mobilized the followers. Management was not institutionalized and decision-making remained concentrated among a small group of individuals.[24] These ENGOs need to institutionalize their managerial structure and improve internal evaluation systems to promote administrative efficiency. As today's environmental problems become more and more sophisticated, the solutions to these problems demand expertise in environmental engineering, chemistry, medical knowledge, biology, meteorology, and other sciences. For example, the controversial and high-profile issue of climate change is a multidisciplinary problem that not only involves natural sciences, but also relates to social sciences including economics, political science, international relations, and sociology. Lack of professional knowledge in these areas will prevent these organizations from playing an active and effective role in influencing governmental decision-makings and

[24]Fu Tao, "Development of Environmental NGOs in China", in Liang Congjie and Yang Dongping (eds.), *The China Environment Yearbook 2005: Crisis and Breakthrough of China's Environment* (Boston: Brill, 2007), p. 307.

setting public agendas. Under such new circumstances, ENGOs need to take a professional rather than a general perspective to cope with certain environmental issues. To better understand public hearings on environmental impact, present evidence more effectively, and work with administrative and bureaucratic systems more successfully, ENGOs have to lift up their professional levels and enhance their capacity building. The Environmental Impact Assessment (EIA) report on the Nu River is a case in point. The report, containing more than 700 pages, touched on various branches of science; so, it is an urgent task for NGO experts to translate the content of the report into scientifically sound and convincing proposals that would lead to professional environmental decision-making.[25] The time of relying on a certain charismatic leader has passed and there is an urgent need for Chinese ENGOs to standardize their operational and managerial procedures and lift up their professional standards. Without a favorable policy condition in China, however, it is very difficult for NGOs to attract fund as well as talented professionals. The *de factor* restrictions from the government, therefore, impose an indirect influence upon the capacity building of ENGOs in China. Despite the large number of ENGOs in the country, they can actually only play very limited roles in supervising the government and setting agendas. The insufficient development of ENGOs hampers China's capability to fight against pollution.

The gap between Chinese and international ENGOs is still significantly large in terms of management structures and professional standards. Unlike many international NGOs that enjoy transnational influence and strong expertise support, Chinese ENGOs are not experienced in the international arena and therefore their perspectives are relatively limited and their actions seldom take on a global outlook.

Furthermore, international ENGOs are more diversified and many of them have special interest area, such as climate change, water pollution, biodiversity protection, acid rain prevention, and animal right protection. In comparison, Chinese ENGOs tend to be more homogeneous. It is easier for many Chinese ENGOs to find problems and file complaints rather than provide solutions. Chinese ENGOs have been expanding exchanges and cooperation with their international counterparts, but the lack of

---

[25] *Ibid.*, p. 308.

English-speaking staff has prevented many of them from frequent contacts with international NGOs or active participation in international talks and forums. Now more and more international negotiations on certain environmental issues invite transnational and indigenous ENGOs to participate, while some international forums and dialogue platforms are firstly initiated or dominated by these ENGOs. However, in many of these occasions, it is difficult to find the deep involvement of Chinese NGOs, whose voices are not often heard in the international arena. Chinese ENGOs still need more support from the government as well as its own endeavor to change its previous charismatic-leader style of operation to a more teamwork-based model, with deepened participation in international dialogues and activities.

Chapter 5

China's Environmental Diplomacy and Soft Power

China's attention to its domestic environmental problems was first triggered by its participation into the international environmental regimes in early 1970s. Its subsequent policy trend in this area has been greatly influenced by the patterns of participation in transnational talks and cooperation on the global environmental commons and the establishment of green norms in the international community.

On the one hand, China was extremely sensitive about the sovereignty issue, which has become paramount when negative externalities cross a country's boundaries and affect the originator's neighbors or the wider international community. One of China's diplomatic guidelines is non-infringement of sovereignty and noninterference of domestic affairs, but collective actions against transnational pollution and ecological deterioration require individual states to relinquish part of its sovereign rights and accept international obligations and surveillance on the basis of established green norms and rules. On the other hand, through frequent personnel and information exchanges with international organizations, governments of foreign countries, NGOs, and environmental activists, Chinese officials have gradually realized that the international environmental cooperation serves the fundamental national interest and is conducive to the elevation of China's international image and soft power in the long run. Through its own rapid economic and social development since the start of reform and opening-up in 1978, China also finds that environmental problems are serious challenges to the communist rule and its solution largely depends on foreign fund, personnel, and technological aid.

CHINA'S ENVIRONMENTAL DIPLOMACY

China started its environmental diplomacy in the early 1970s when the People's Republic of China (PRC) replaced Taiwan (Republic of China) to take the legitimate seat in the United Nations and PRC warmed relations with western countries, especially the United States. At that time, an urgent demand for the PRC that had just returned to the international community was to set up a special organization in preparation for the upcoming United Nations Conference on the Human Environment (UNCHE) in June 1972. The Chinese people were still in the furor of Cultural Revolution, and many existing governmental departments had lost their original socioeconomic functions. Environmental protection was completely a new issue to China, with no corresponding governmental organs to take charge. As one of the first important international conferences after PRC's return to the United Nations, the UNCHE in Stockholm aroused extremely intensive attention from the Chinese government, which regarded it as a best diplomatic chance to make a show of socialist China to the western world.

In 1971, a leading panel on environmental protection was set up informally under the State Council to make preparations for the 1972 UNCHE. The panel that was first established in direct and urgent response to an impending diplomatic requirement became the predecessor of the future environmental department in China. China did not play a particularly constructive role at UNCHE. Like developing countries in general, it assigned principal responsibility for pollution control to the advanced industrialized countries and defended the right of developing countries to exploit their own resources without external interference.[1] China was an inactive

[1]Lester Ross, "China: Environmental Protection, Domestic Policy Trends, Patterns of Participation in Regimes and Compliance with International Norms", in Richard Louis Edmonds (ed.), *Managing the Chinese Environment* (Oxford: Oxford University Press, 1998), p. 87.

participant of this conference, and its delegation even tried to inject such tangential and divisive issues as the Vietnam War and nuclear testing into the Conference declaration. China was a "laggard" participant in this international regime, avoiding international obligations by shunning treaty commitments or exhibiting a disdainful attitude toward compliance obligations.[2]

In the eyes of not only China, but also the rest of the world, UNCHE is regarded as the origin of international environmental diplomacy. UNCHE had two direct impacts upon China: providing environmental awareness for China and giving impetus to China for environmental policy making. Through exchanges and talks among delegations from different countries, Chinese officials came to realize that environmental issues that they once ignored may carry strategic significance and become vital for a nation's existence and development. Similar experiences whereafter proved that such international socialization process helped China to reconstruct its national interest on the premise of new information and knowledge.

From the traditional perspective of real politics during the Cold War, China often feared that western industrialized countries would make use of environmental concerns to curtail their economic potential and criticize their policy. Such mindset even lasts until today and still has great influence upon China's stance on some international environmental issues such as climate change. Yet, China's original hostile attitude toward western countries' persistence on green issues has been softened by large margins due to the long-time contact, exchanges, and learning process on many international occasions. Some officials attending the UNCHE and other early international conferences on environmental issues became very influential officials working for China's environmental department in the future. Qu Geping, who later became the chief administrator of the National Environmental Protection Bureau in 1980s and won some international prizes for his outstanding contributions to China's green causes, was a member of the delegation to the UNCHE in 1972.

UNCHE not only marked the start of China's environmental diplomacy and domestic environmental governance in the modern sense, but also was widely regarded as the origin of international environmental diplomacy. International environmental agreements, addressing external effects both

---

[2] *Ibid.*

on the global commons and within another country's frontier, have increased in number and entered into force more rapidly since UNCHE.[3] China began to play an increasingly active role in the international environmental diplomacy, participating in more international institutions and expanding bilateral and multilateral exchanges in this area. In 1973, China took part in activities initiated by the United Nations Environment Programme (UNEP), which joined hands with the Chinese government to set up a research and training center on desertification prevention in China in 1987.[4] Since 1979, UNEP sponsored a number of symposiums and training courses on environment impact assessment, ecological protection in rural areas, climate change, water pollution prevention, and biodiversity protection.

The World Wide Fund for Nature (WWF), the first international conservation organization invited to work in China, has been active here since 1980, when its experts arrived to work with the Chinese scientists on giant panda conservation. In 1996, WWF opened its Beijing office with nine staff working on four programs: Giant Panda Conservation, Energy Efficiency, Forests, and Environmental Education.[5] It is the State Forestry Administration rather than the State Environmental Protection Administration (SEPA) that has the closest ties with WWF. Today, WWF China Program Office has about 80 staff working on over 40 projects, from restoring the Yangtze river wetlands to environmental education and panda conservation. In addition to the Beijing office, WWF now also has six field offices in Changsha, Hunan province; Wuhan, Hubei province; Kunming, Yunnan province; Xi'an, Shaanxi province; Chengdu, Sichuan province; and Lhasa, Tibet.[6] In 1986, China's National Environmental Protection Bureau set up regular ties with the World Bank, which started to offer loans to some anti-pollution projects in the country.

---

[3] Peter M. Haas and Jan Sundgren, "Evolving International Environmental Law: Changing Practices of National Sovereignty", in Nazli Choucri (ed.), *Global Accord: Environmental Challenges and International Responses* (Cambridge, MA: The MIT Press, 1993), pp. 405, 410–411.

[4] China Environment Yearbook Editing Committee (ed.), *China Environment Yearbook (1990)* (Beijing: China Environmental Science Press, 1990), p. 209.

[5] "About WWF China", WWF China, http://www.wwfchina.org/english/loca.php?loca=60.

[6] *Ibid.*

In 1996, China became a state member of the International Union for Conservation of Nature (IUCN), a global union of members devoted to balancing human needs with conservation objectives. As early as 1980s, IUCN started to be actively involved in a range of conservation work in China, including species assessments, protected area management and World Heritage, environmental law, and environmental economics. The IUCN China Liaison Office was established in January 2003, and a Memorandum of Understanding was signed with the State Forestry Administration (SFA) of China in 2004. IUCN has also been an active member of the China Council for International Cooperation on Environment and Development (CCICED) since its inception, and IUCN experts have participated in a number of its Task Forces.[7] The CCICED, established with financial support from Canada in 1992 and once chaired by then Vice-Premier Wen Jiabao, provides a formal avenue for the communication of international advice to the Chinese government.

China's increasing involvement in international environmental cooperation and conferences pushed the country to make more effort domestically to cope with environmental issues. China's first set of environmental regulations (except for a few occupational health and safety norms adopted from Soviet models), "Some regulations on protecting and improving the environment," was circulated in 1973.[8] From 1972, China stated the need to incorporate the environment in the national planning process, although environmental protection did not receive treatment as a separate chapter in such national plans until the 6th Five-Year Plan (1980–1985).[9]

During 1980s when the international environmental movement and diplomacy became more vibrant and China put its policy focus on reform and opening-up, the Chinese government carried out frequent contacts and exchanges with foreign governments on environmental protection. In 1989 alone, the Chinese government sent out 80 delegations to participate

---

[7] "About IUCN in China", IUCN, http://cms.iucn.org/about/union/secretariat/offices/asia/asia_where_work/china/iucnchina_about/index.cfm.

[8] Lester Ross, "China: Environmental Protection, Domestic Policy Trends, Patterns of Participation in Regimes and Compliance with International Norms", in Richard Louis Edmonds (ed.), *Managing the Chinese Environment* (Oxford: Oxford University Press, 1998), p. 88.

[9] *Ibid.*

in international environmental activities,[10] including the first meeting of the parties to the Montreal Protocol on Substances that Deplete the Ozone Layer, the London Conference on Saving the Ozone Layer, Intergovernmental Panel on Climate Change (IPCC) conference, and the International Conference on Hazardous Waste Management. Protection of the ozone layer and implementation of the Montreal Protocol became the key environmental topic of that year, when China expressed its strong support for the joint action to save the ozone layer and proposed the establishment of an international fund in this area. In 1989, the Ministry of Foreign Affairs began to train environmental specialists in its international department and treaties and law department. In the same year, former Chairman of State Commission on Environmental Protection (SCEP) Song Jian articulated the term environmental diplomacy (*huanjing waijiao*) for the first time at the 16th meeting of SCEP.

During 1990s, China's environmental exchanges with the outside world became more active and frequent. In 1995, the National Environmental Protection Bureau sent out 250 delegations and received 80 visiting delegations.[11] In the same year, China participated in some important international conferences on environmental protection, including the Conference of Parties (COP) 1 of the United Nations Framework Convention on Climate Change (UNFCCC) in Berlin, COP 3 of the Basel Convention on the Control of Transboundary Movements of Hazardous Wastes and Their Disposal in Geneva, COP 2 of the Convention on Biological Diversity in Jakarta, Vienna Meeting of the Parties to the Montreal Protocol on Substances that Deplete the Ozone Layer, the 18th Meeting of the UNEP Governing Council in Nairobi, and the Ministerial-Level Conference on Environment and Development in Asia and the Pacific in Bangkok. By the end of 1995, the National Environmental Protection Bureau altogether received foreign donation worth US$ 210.6 million.[12]

---

[10]China Environment Yearbook Editing Committee (ed.), *China Environment Yearbook (1990)* (Beijing: China Environmental Science Press, 1990), p. 210.

[11]China Environment Yearbook Editing Committee (ed.), *China Environment Yearbook (1996)* (Beijing: China Environmental Science Press, 1996), p. 231.

[12]*Ibid.*, p. 237.

From the latter part of the decade, China gradually integrated international environmental cooperation into the highest level of diplomatic agendas. In 1999, the 11th Meeting of Parties to the Montreal Protocol on Substances that Deplete the Ozone Layer and the 5th Meeting of COP to the Vienna Convention was convened in Beijing. Jiang Zemin, then China's president, addressed the conference that "China takes a responsible attitude towards global environmental protection, and has been making persistent efforts to tackle domestic environmental problems".[13] During Jiang's visit to Japan in 1998 and Premier Zhu Rongji's visit to the United States and Canada, environmental protection became the important topic of the bilateral talks with senior officials from the State Environmental Protection Administration as entourage.[14] Local environmental departments also paid more attention to foreign exchanges and cooperation. From 1996 to 2000, one-third of all provincial-level environmental protection bureaus set up special branches in charge of international cooperation.[15]

China continued to expand the transnational economic cooperation on environmental projects, attracting a foreign fund of US$ 627 million in this regard by the end of 2002.[16] After 2000, such trend of active involvement in international environmental diplomacy continued, with more multi- and bilateral-environmental treaties signed by China and more environmental conferences and talks attended by Chinese leaders. Important environmental conferences attended by China during this period include the Meeting of Parties to the Montreal Protocol, Conference of Parties to the Kyoto Protocol, the Ministerial-Level Conference on Environment and Development in Asia and the Pacific, Northeast Asian Conference on Environmental Cooperation (NEAC), and the Conference of the Parties to the Stockholm Convention on Persistent Organic Pollutants (POPs).

---

[13]"Jiang Zemin Addresses International Conference on Ozone Layer Protection", *People's Daily Online* (3, December 1999), http://english.people.com.cn/english/199912/03/eng19991203A104.html.

[14]China Environment Yearbook Editing Committee (ed.), *China Environment Yearbook (2001)* (Beijing: China Environmental Science Press, 2001), p. 232.

[15]*Ibid.*

[16]*Ibid.*, p. 237.

Since China's attendance to UNCHE in 1972, its patterns of participation in international environmental regimes have changed dramatically. Generally speaking, China has over time become more willing to join such institutions and accept treaties in this area than before. Regarding this, it may be noted that the Taiwan (Republic of China) government, which previously occupied China's seat in the United Nations and specialized international organizations, was not an active participant in international environmental law at that time.[17] Moreover, China's adherence to international obligations has become more pronounced in recent years as international environmental diplomacy has accelerated and as China's own capacity to participate has increased.[18]

Although China has always been cautious about international regimes' possible infringement on national sovereignty, it gradually cultivated a more open and tolerant attitude toward new environmental treaties and relevant obligations accrued to it. The Convention on Biological Diversity, adopted at the 1992 United Nations Conference on Environment and Development (UNCED) in Brazil, imposed considerable constraints on development in undeveloped areas to preserve biological diversity within a jurisdiction. China, with adherence to the principle that developing countries' right to develop should be respected in the process of environmental protection, surprisingly gave strong endorsement to the Convention on Biological Diversity and became the first major state (and 5th overall) to ratify the convention. China's concern that its economic growth may be hampered by environmental commitment is there, however. In dealing with climate change and other green issues with broad constraint upon economic development, China took a less prominent role and refused to shoulder the same compulsory obligations as industrialized nations. With the exemption of mandatory emission cuttings, China gladly ratified both the Framework Convention on Climate Change and the Kyoto Protocol, the two milestone treaties on human's joint effort against global warming.[19]

[17]Lester Ross, "China: Environmental Protection, Domestic Policy Trends, Patterns of Participation in Regimes and Compliance with International Norms", in *Managing the Chinese Environment* (Oxford: Oxford University Press, 1998), p. 92.

[18]*Ibid*., p. 88.

[19]For details of China's participation in international cooperation against climate change, please refer to Chap. 9.

SOFT POWER AND INSTITUTIONAL PARTICIPATION

Like other members of the international community, China's environmental diplomacy and its participation into the global environmental regimes have great influence upon its soft power. There are many definitions of soft power. According to Professor Joseph S. Nye, Jr., who first developed the concept of "soft power" in *Bound to Lead* published in 1990, soft power "is the ability to get what you want through attraction rather than coercion or payments."[20] It arises from the attractiveness of a country's culture, political ideas, and policies.[21] Soft power uses a different type of currency (not force, not money) to engender cooperation — an attraction to shared values and the justness and duty of contributing to the achievement of those values.[22] Hence, if a country's foreign policy and practices abide by the commonly recognized international norms and win the support and respect from others, its soft power will be fortified. Moreover, soft power is in play if a country can effectively exercise its influence to set the agenda in international affairs, have its positions and argument accepted by the majority of the international society, shape other states' preferences, have the ability to offer suggestions to solve problems, delegitimize others' positions or advocacies, and build a coalition through noncoercive means.

China, with sizzling economic growth for more than two decades, has witnessed its hard power greatly enhanced in the international arena, processing consolidated material strength and increased confidence to fulfill its strategic goals and expand its global influences. Correspondingly, its soft power has also been growing remarkably thanks to domestic economic miracles, charming of its unique culture and history as well as successful and seasoned foreign policy. The growth of China's soft power comes not only from the domestic front, but also from its participation in various international regimes.

---

[20]Joseph S. Nye, Jr., *Soft Power: The Means to Success in World Politics* (New York: Public Affairs, 2004), preface x.
[21]*Ibid.*
[22]*Ibid.*, p. 7.

Environmental protection has become an important issue of international focus since 1970s. The green movement since then in the global scale called for international joint actions to protect the ecological commons. Due to its large population, rapid economic growth, and heavy reliance on fossil fuels, China played an especially important role in the international environmental talks as well as the formation of environmental institutions. China has already been the world's largest sulfur dioxide emitter for years and some figures show that it has replaced the United States to be the world's largest greenhouse gas (GHG) emitter in 2007.[23] China estimated its GHG emissions reached 6100 million tons CO_2 equivalent in 2004, accounting for 12.45% of the world's total 49,000 million tons.[24] Whether it can maximize its national interest on green issues and gain diplomatic support from most others are closely related to how it uses soft power to attract others to its side and fulfill a win–win situation that benefits all. Meanwhile, its acceptance of international environmental norms and support of the international climate accords have enhanced its soft power and gained respect from others. Active participation in international environmental negotiations and institutions has been an important part of China's diplomatic activities since early 1990s, which effectively promoted China's international image, elevating its soft power.

Institution Participation Enhances Soft Power

Voluntary participation in major international institutions and compliance of established rules in the international community are essential for the soft power maintenance. International institutions are "the rules that govern elements of world politics and the organizations that help implement those

[23]Netherlands Environmental Assessment Agency, "China Now No. 1 in CO_2 Emissions; USA in Second Position", Netherlands Environmental Assessment Agency Press Releases 2007 (19, June 2007), http://www.mnp.nl/en/dossiers/Climatechange/moreinfo/Chinanowno1inCO2emissionsUSAinsecondposition.html.

[24]The statistics are respectively from National Development and Reform Commission (NDRC), "China's National Climate Change Program", China Climate Change Info-Net, June 2007, p. 6, http://www.ccchina.gov.cn/WebSite/CCChina/UpFile/File188.pdf and Intergovernmental Panel on Climate Change (IPCC) Working Group III Report, Mitigation of Climate Change (2007), p. 103, http://www.ipcc.ch/ipccreports/ar4-wg3.htm.

rules".[25] Although the international society is still described as anarchical with world politics driven by the traditional exercise of state power, international institutions are increasingly important and no longer marginal. In reality, even the most powerful states were relying increasingly on international institutions — not so much on the UN as other organizations and regimes that set rules and standards to govern specific sets of activities.[26] Institutions create the capability for the states to cooperate in mutually beneficial ways by reducing the costs of making and enforcing agreements — what economists refer to as "transaction costs."[27] International institutions become extremely important when sovereign states nowadays are facing more global problems that need joint effort and policy coordination. These urgent problems include terrorism, transnational crimes, global warming, trade protectionism, hot money flow, increasing epidemic diseases, and overuse of common-pool resources. Without necessary institutions, self-interested sovereign states face the prisoners' dilemma and a few states are willing to take voluntary actions for collective benefit of the international society. Certain institutions and rules have to be set up to provide incentives and disincentives for individual states in a bid to form collective actions.

By reducing the uncertainty of enforcing agreements, international institutions help states achieve collective gains.[28] Meanwhile, participating in international institutions means acceptance of certain type of commonly-recognized norms, rules, or even culture. Joining the Non-Proliferation Treaty (NPT) or the Convention on Biological Diversity is the indication that a state recognizes and is willing to follow the international norms of limiting the spread of nuclear weapons or conservation and sustainable use of biological diversity. Therefore, a state's stances over various kinds of international institutions are extremely important for its soft power expansion, because this directly affects its moral attraction in the eyes of other states that value the cooperation in the frameworks of these

---

[25] Robert O. Keohane, "International Institutions: Can Interdependence Work?", *Foreign Policy*, Vol. 110 (1998), p. 82.
[26] *Ibid.*, p. 84.
[27] *Ibid.*, p. 86.
[28] *Ibid.*

institutions. The soft power of a country rests primarily on three resources: its culture, its political values, and its foreign policies.[29] Whether to join an international institution is not only a part of a country's foreign policies, but also an indicator of its political values.

China's Active Participation in International Institutions

As a permanent member of the UN Security Council, China is a signatory to a large number of international treaties and a member of many influential international organizations. In the field of disarmament and arms control, China signed almost all the major international accords including the NPT, the Comprehensive Nuclear-Test-Ban Treaty (CTBT), the Convention on the Banning of Chemical Weapons, the Convention on the Banning of Biological Weapons, and the Convention on Certain Conventional Weapons; in the field of environmental protection, China joined more than 50 international environmental treaties including the well-known UN Framework Convention on Climate Change, the Montreal Protocol, the Kyoto Protocol, the Convention on Biological Diversity, the UN Convention to Combat Desertification, the Rotterdam Convention on the Prior Informed Consent Procedure for Certain Hazardous Chemicals and Pesticides in International Trade, and the Stockholm Convention on Persistent Organic Pollutants. Despite its deficient protection of intellectual property right domestically, China is a member of the World Intellectual Property Organization, the Berne Convention for the Protection of Literary and Artistic Works, the Madrid Agreement Concerning the International Registration of Marks, the Universal Copyright Convention, and the Agreement on Trade Related Aspects of Intellectual Property Rights (TRIPS) administered by the World Trade Organization. Although the United States often attacks China's human rights record, China ratified more international treaties on human rights protection than the United States. As the world's second largest greenhouse gas emitter for a long time, China's participation in the international climate regime and the Kyoto Protocol shows how the country

---

[29]Joseph S. Nye, Jr., *Soft Power: The Means to Success in World Politics* (New York: Public Affairs, 2004), p. 11.

gained its soft power through joining the international institution that seeks common interest of the globe. It also means that if China fails to make further commitment as expected by the international community in the next round of climate talks, its soft power will be probably weakened.

Chapter

6

Case Study: Songhua River Toxic Spill Crisis

On 13, November 2005, an explosion jolted a state-owned petrochemical plant in Jilin City, about 380 kilometers (230 miles) up the Songhua River from Harbin, the densely-populated capital city of northeast China's Heilongjiang Province. The explosion, killing six and injuring dozens, later triggered one of the most serious toxic spills in China since the start of the reform and opening-up in 1978. The horrible blasts later created an 80 kilometer long toxic slick in the Songhua River, a tributary of the Amur River. This crisis not only posed a threat to the safety of drinking water for residents along the river, but also aroused social panic, public anger, and even transnational disputes. To appease the widespread discontent about officials' breach of duty throughout the crisis, once-meritorious Minister Xie Zhenhua of the State Environmental Protection Administration (SEPA) was pressured to resign, making him the highest-ranking environmental official to step down as a result of pollution crisis. Since the Songhua River pollution in 2005, other mishandled pollution crises endangering public health and safety continued to happen in different parts of China. The Songhu River toxic spill crisis became a typical case that reflected the faulty management in China's environmental governance. A review of the whole crisis management process will help people to understand that much of China's pollution actually should be attributed to China's officialdom culture, governmental structure and other governance factors instead of the expanding economy and large population.

REVIEW OF THE SONGHUA RIVER POLLUTION CRISIS

At 1:45 p.m. on 13, November 2005, an explosion occurred in the benzene processing section of the No. 101 Petrochemical Plant operated by the Jilin Petrochemical Company, a subsidiary of PetroChina, one of China's largest state-owned oil giants. A vast mushroom cloud was reportedly seen afterward and the whole Jilin City was enshrouded by the chemical haze. A series of explosions in the chemical plant killed six, injured dozens, and caused the evacuation of tens of thousands of residents.

Jilin Petrochemical provided an explanation of the explosion's cause in a press release issued that evening: a mishandled blockage in a chemical processing tower with the designation P102 produced a fire in that unit, which in turn triggered an explosion in the nitrifying unit used for aniline processing.[1] The front-page coverage of the explosion in the Jilin Petrochemical Plant immediately drew people's attention to the possible pollution, especially air pollution thereafter. According to a report written by Xinhua News Agency, the official mouthpiece, on 15 November, the relevant environmental supervision department launched an emergency pollution monitoring on spot, having found benzene and phenol in the air above the chemical plant. The report reassured local residents by saying that "these chemicals disperse quickly fortunately and people can return home and resume their normal lives after a short evacuation period." During the period from 13 to 21 November, local people's life gradually came back to normal with reports relating to the explosion dwindling every day.

Just as the chemical blast almost lost its front-page position in the local newspaper, residents of Harbin — a downstream city of Jilin City along the Songhua River — started to spread the rumor that the Songhua

---

[1]Qie Jianrong, "Warnings and Lessons from the Songhua River Toxic Spill Crisis", in Liang Congjie and Yang Dongping (eds.), *The China Environment Yearbook 2005: Crisis and Breakthrough of China's Environment* (Boston: Brill, 2007), p. 20.

River, from which they drew the drinking water, had probably been contaminated by deadly chemicals leaked out from the factory explosion. As no official statement was announced to refute those rumors, residents in Harbin started to stockpile water in whatever containers they found at home. The most ironic event came on 21 November when the Municipal Government of Harbin made an announcement that the tap water supply would be cut off for four days starting from 12:00 a.m., 22 November, for an overall check and maintenance of the city's pipeline network. It was an obvious lie to the public when most residents had already become suspicious about their urban water quality due to the widespread "rumor." As more and more people came to know the truth, the Harbin government had to announce later that day again that the water in the Songhua River had been seriously polluted by the chemicals from the Jilin petrochemical plant's explosion. This announcement officially concluded the 8 days of quietness after the blast in Jilin, and people not only in Harbin but also all over China began to realize that a real crisis had arrived. How the government was to supply drinking water for the more than three million urban residents in Harbin became a headache. According to China's official report, the explosion severely polluted the Songhua River, with an estimated 100 tons of pollutants containing benzene and nitrobenzene entering into the river.[2] Exposure to benzene reduces white blood cell count and is linked to leukemia. The water contamination actually took place immediately after the blast on 13 November, and Harbin was only one of the places along the river suffering from the toxic slick, which first passed through several counties and cities in Jilin Province, including Songyuan, and then entered Heilongjiang Province, traversing Harbin, Jiamusi City, and other counties. Dead fish could be seen along the banks of the Songhua upstream from Harbin. The contamination soon had an international impact when the slick converged into the Amur River at the mouth of the Songhua River on the border between China and Russia. It passed by the Jewish Autonomous Oblast in Russia, and then entered the Russian region of Khabarovsk Krai in the Russian Far East, passing through the cities of Khabarovsk and Komsomolsk-on-Amur before exiting into the Strait

[2]"China Pledges to Minimize Impact of River Pollution on Russia", *Xinhua News* (24, November 2005).

of Tartary, itself a bridge between the Sea of Okhotsk and the Sea of Japan portions of the Pacific Ocean.

After the government officially acknowledged the Songhua River contamination, the information release became transparent again and responding actions were launched effectively. On 23 November, Harbin residents began to receive water from fire trucks, and began voluntary evacuation. Schools in Harbin were closed for 1 week. On 24 November, the State Council Information Office held a special press conference on the toxic spill issue, in which Deputy Chief Zhang Lijun of the SEPA told the press that a daily briefing system on the crisis would start from 25 November.

Water supply in Harbin was resumed in the evening of 27 November, and on 29 November, the SEPA sent an expert panel to Harbin to further investigate the contamination and work out plans of weeding out the pollutants. On 26 December, the SEPA publicized the latest test results along the Songhua River, showing that concentrations of nitrobenzene and benzene at all monitoring stations were under national limits for these chemicals. As China adopted stricter environmental standards for these chemicals than Russia, the Russians were appeased by these results. To avoid the Sino-Russian relations from being affected by the water crisis, Chinese President Hu Jintao kept a regular contact with his counterpart in Russia. China agreed to offer water testing and purifying materials, including 1000 tons of activated carbon to Russia.[3] In response to the request from Russia, China started to construct a diversionary dam on the Fuyuan waterway in Heilongjiang Province on 16 December to prevent the spread of pollution.

WHO SHOULD BE RESPONSIBLE?

The Songhua River pollution crisis caused social panic, inconvenience as well as huge economic losses. The initial announcement of water stoppages led to panic buying of water and food, exhausting supermarket supplies, and sending prices soaring. The Harbin City was full of ridiculously large

[3] "Second Batch of Water Purifying Materials offered to Russia", ChinaView (16, December 2005), http://news.xinhuanet.com/english/2005-12/16/content_3932826.htm.

queues, and people were buying water in massive quantities. The accident forced Harbin to suspend water supply from 22 to 27 November because of contamination. During that period, 29 industrial enterprises in the city stopped production and 23 reduced production, leading to a daily loss in output value of nearly 100 million yuan.[4] Chen Jinsong, a researcher with the development and research center of Heilongjiang provincial government, estimated that at present the direct loss brought about by the accident might amount to some 1.5 billion yuan, according to a report carried by *China Economic Times*.[5] As the disaster happened during winter, Harbin, China's ice capital famous for the unique landscape and colorful sports activities in the ice and snow, was estimated to lose 5 billion yuan from its tourism income in the season. Besides, China's international image got damaged from the disaster due to the transnational harm to Russians and the government's slowness and incredibility in handling the crisis at the initial stage.

Just like dealing with similar incidents caused by dereliction of duty before, the Chinese government had to punish some officials to show its strong determination to safeguard people's safety and welfare. SEPA, being in charge of China's environmental protection affairs, became the first to be blamed for the pollution. Xie Zhenhua, then chief of SEPA, resigned soon after the Songhua River contamination, and was replaced by Zhou Shengxian, former director of the State Forestry Administration. Xie, the highest-ranking environment official to leave a job in such a case, had 30 years of working experience in the area of environmental protection. In 2003, he was awarded the Sasakawa Environment Prize by the United Nations for his unremitting efforts in the green area during the past two decades. Kofi Annan, then UN Secretary-General, said at the ceremony that Xie "has shown great zeal and imagination in promoting the concept of sustainable development in China".[6]

Many people in the circle thought the punishment was unfair to Xie, who was a thousand miles away from Jilin when the explosion took place and could only play a very limited role in handling the issue due to SEPA's

---

[4] Shao Da, "Paying the Bill for Songhua River Pollution", CHINA.ORG.CN (5, December 2005), http://www.china.org.cn/english/2005/Dec/150917.htm.
[5] *Ibid.*
[6] "Xie Zhenhua Awarded Sasakawa Prize", *Xinhua News* (21, November 2003).

long-time weakness in local pollution control. Local vested interest prevented SEPA from knowing the truth and the local branch of SEPA was under great pressure from other administrative departments in the province at the initial stage of the crisis. No local official in Jilin Province was forced to step down after the pollution, while Yu Li, head of the Jilin branch of China National Petroleum Corporation (CNPC), was dismissed for his responsibility for the toxic spill. Unlike CNPC's deadly blowout accident in 2003 in Chongqing,[7] this time the senior management of CNPC's headquarters in Beijing was not affected at all.

Actually even today, it is not clear as to who was responsible for the 8 days (13–21 November) wasted after the chemical blast. According to the statement made by SEPA Deputy Chief Wang Yuqing in the National Environmental Pollution Accident Videoconference on 1, December 2005, in the few days after the eruption of the Songhua River toxic spill crisis, i.e. from the 14th to 17th of November, the SEPA had received no information from Jilin Province's environmental agencies concerning this severe pollution accident, and thus the SEPA missed the best opportunity to contain the crisis.[8] This announcement, however, seemed contradictory to some local environment officials' statement saying that during that period, relevant data concerning water pollution had been collected by the local environment monitoring stations without any delay. It was not understandable for all the monitoring stations under the Jilin Provincial Environmental Protection Bureau not to pay any attention to the water quality after such a huge blast took place in a chemical plant along the Songhua River.

There was no direct incentive for SEPA to hide the pollution fact reported by its local branches in Jilin either, because it was not SEPA's responsibility to prevent the toxic chemicals from leaking into the Songhua

[7]The gas blowout, killing 243 people and injuring over 4000 in southwest China's Chongqing Municipality, took place on 23, December 2003 at a natural gas field in Chongqing's Kaixian County operated by China National Petroleum Corporation (CNPC), when a well burst and released a high concentration of natural gas and sulfurated hydrogen. CNPC General Manager Ma Fucai resigned later with an acknowledgment of breach of duty.

[8]Qie Jianrong, "Warnings and Lessons from the Songhua River Toxic Spill Crisis", in Liang Congjie and Yang Dongping (eds.), *The China Environment Yearbook 2005: Crisis and Breakthrough of China's Environment* (Boston: Brill, 2007), p. 22.

River. It was possible, however, for the local government to cover the truth because of the mindset of *paluan* (fear of social instability) in China's officialdom. Administrative officials possibly were more concerned about the widespread social panic and fury should the public come to know the facts about this severe water pollution. According to China's political culture, such high-profile safety accident will affect the future promotion of the mayor, governor, or even the party secretary in that province; so, the first choice for many of them was to hold back the frightening report and make remedial actions tacitly to minimize the social impact. Such actions may backfire if the event becomes uncontrollable and the solution demands the involvement of more departments and areas.

China's initial failure in the fight against epidemic SARS and some officials' intentional cover-up of SARS cases in 2003 were other typical examples showing this commonly-seen political practice in China. Immediately after the blast, officials in Jilin Province showed their eagerness to minimize the social impact of this safety accident to the lowest level. At the press conference held on the morning of 14 November, a vice mayor of Jilin City told the press that expert analysis found no evidence of significant pollution resulting from explosion; no toxic gases or contaminated water was detected at the site or in the surrounding areas.[9] Even on 22 November, 9 days after the blast, a director from the Jilin Petrochemical Company still announced that no water contamination was caused by the explosion because the plant had its own wastewater treatment facilities. Only after one day, when they found it too difficult to cover up the truth, did both the SEPA and Jilin Provincial Government officially acknowledge the serious water contamination in the Songhua River.

Some reports later showed that local environment officials had been silenced and excluded from the investigation after the blast. According to the report entitled "Eight Days — Time Wasted" in *China Newsweek*, Zhao Jing, Director of Changchun Environment Monitoring Station, got to know from her schoolmate in the Jilin Environmental Protection Bureau that the staff in the bureau were not allowed to get involved in the monitoring.[10] Instead, the Ninth Mineral Institute of CNPC was appointed

---

[9] *Ibid.*
[10] *Ibid.*, p. 23.

to conduct the pollution monitoring. As a scapegoat, Xie reportedly wept with his outspoken colleague Pan Yue, deputy chief of SEPA, in the office building of SEPA upon the compelled resignation.

GOVERNANCE PROBLEMS REFLECTED

The Songhua River pollution crisis revealed some serious problems hidden in China's environmental governance systems. Firstly, SEPA and its local branches lack sufficient authority to enforce environment regulations and to supervise and monitor pollutions. The SEPA could not fulfill its responsibilities of pollution control without adequate administrative power in hand. Before 1998 when the SEPA was promoted to the ministerial level, there was another Environmental Protection Commission (EPC) under the State Council that enjoyed more power and independence than SEPA. The EPC had long time been headed by State Councillor Song Jian whose rank equaled vice premier, and included ministers of relevant departments and officials of provincial governments. The EPC was actually a vice-premier-rank inter-agency organization that took charge of all the work related to environmental protection. After 1998, SEPA had very limited pollution control power, which was also shared and scattered among other ministries and departments such as the Ministry of Water Resources, the Ministry of Agriculture, the Ministry of Construction, the Ministry of Land and Resources. In terms of water pollution in major rivers and lakes, SEPA only had the right to monitor and collect data, while the Ministry of Water Resources had greater power in water management. In the Songhua River region, the "Songliao Committee" under the Ministry of Water Resources sometimes has clashes with the monitoring station operated by SEPA.

Similar complicated situations exist along the Huaihe River, the Yangtze River, and other big rivers and lakes. Besides, local environmental protection bureaus (EPBs) are under the dual leadership of its superior environmental protection department as well as the local government/party committee. EPBs are more dependent upon local governments, which provide them with annual budgetary funds, personnel, cars, office buildings, and even employee housing. SEPA leaders, however, are based in Beijing, the faraway capital, and thus have much fewer contacts with the local branches than the local government and party bosses. Such an attachment to local leadership has prevented EPBs from functioning efficiently against pollution and ecological degradation.

The Songhua River toxic spill also reflected the long-time accountability problem existing in China's officialdom. The Chinese government has realized that when something goes wrong in the governance, certain officials have to be responsible for the mishandling and get punished. Yet, the problem now is that in many cases, the officials being punished are not those who should be most responsible for the faulty activities. The after-the-event investigation is often as opaque as the information released during the crisis, with the public having no idea about whether the real culprits have received due punishment or had a scapegoat been used instead. Sometimes, it is the whole system or the political culture rather than individuals that should be most responsible for many environmental disasters.

Moreover, the crisis may reveal that some staff members in local environmental protection departments lack special expertise to fulfill their pollution monitoring and supervision duties. It is quite common in many parts of China that staffing local governmental agencies relies on political and social connections rather than expertise. In the Environmental Protection Bureau of Jieshou City, Anhui Province, in 2006, only 5 of the total 59 staff members were professionals in the environmental area.[11]

Finally, the Songhua River pollution disaster also exposed China's weakness in protecting public interest through litigation means, as well as its hesitation in severely punishing those state-owned giants for industrial accidents. Although the country frequently witnesses serious pollution disasters that cause huge economic losses and even endanger people's safety, it is still not easy for one to act as a public interest representative and seek compensation through lawsuit from the government or the state-owned corporations for ecological losses. If such public compensations could be gained at court, the state-owned CNPC, the culprit of this water pollution

[11]"Local Environmental Protection Departments Not Well Equipped", (*difang huanbao bumen tige tairuo*), *Economic Information Daily* (29, September 2006), http://news.xinhuanet.com/fortune/2006-09/29/content_5152829.htm.

crisis, would have been bankrupted a year ago when its blowout accident killed 243 people near the Chongqing City. During the Songhua River pollution, the water contamination, water stoppage, and damage of reputation to the local tourism industry affected millions of people and thousands of factories and businesses in Jilin and Heilongjiang Provinces. In a western country, the economic compensation for the public interest could be an astronomical figure and lead to the immediate bankruptcy of the troublemaker. Due to the lack of such litigation traditions, both state-owned industrial giants and the government tend to be bolder when handling severe pollution accidents.

Chapter 7

Case Study: Taihu Algae Crisis

At the end of May 2007, a severe algae outbreak in the Taihu Lake, China's third largest freshwater lake, rendered tap water foul-smelling and undrinkable over the course of a week for half of the 2.3 million residents in Wuxi, a wealthy city in eastern China's Jiangsu Province. This worst-ever water crisis in Wuxi immediately caused social panic, touching off local residents' panic buying of bottled water in the supermarkets and even class suspension of some local colleges.

By mid-June, a second algae bloom appeared in the central-western and northern parts of the lake, covering 800 square kilometers and causing widespread concern in cities around the lake. About 30 million people rely on Taihu Lake for drinking water, including nine cities in the provinces of Jiangsu and Zhejiang as well as Shanghai Municipality.

High levels of nitrogen and phosphorus are believed to be the major causes of algae blooms, which develop in water that is rich in nutrients. Apart from Taihu Lake, blue-green algae outbreaks have been reported in Chaohu Lake and southwestern Dianchi Lake since May 2007, and have threatened the local tap water supply.

The blue-green algae outbreak in large lakes was only one symbol reflecting China's quickly-deteriorating environment and ecosystems in recent years, when the country was turning itself into a "world workshop" with the economy surging at a double-digit growth rate per annum.

Environmental pollution is a global issue that most fast-growing economies have to face, relating to many factors such as economic structure, technological level, political systems, governance capacity, institutional building as well as public awareness and social participation. In China, however, the increasing environmental crisis is more like a result of its unique political and economic institutions, only showing the weak capacity of the government to handle those "growing pains." The case study of Jiangsu's environmental problems shows improvement of governance quality, and style is a better solution to China's pollution, rather than simply enhancing fund and technology investment.

FAILURE OF 16-YEAR EFFORT TO CLEAN-UP TAIHU LAKE

The Taihu Lake, once well-known for its limpid lake water and beautiful scenery in ancient China, has become one of the most polluted freshwater lakes in China since 1980. From 1991, China's central government and localities started clean-up projects against pollution in Taihu Lake with huge fund and human resource input. Throughout the year 1998, the State Environmental Protection Administration (SEPA), together with other ministries and provincial-level governments of Jiangsu, Zhejiang, and Shanghai, launched the "Zero O'clock Action" (*lingdian xingdong*) to check and close pollution sources around the Taihu Lake with the deadline on zero o'clock of 1, January 1999. Just like another "Zero O'clock Action" against Huaihe River in 1997, this campaign in Taihu proved to be ineffective with nitrogen and phosphorous content maintaining at high levels. Although later continuous efforts were made and the Chinese government even made pollution control in this area a key program in the 10th Five-Year Plan (2001–2005), no substantial change took place and the water quality even worsened. According to SEPA's survey in 2005, the water quality of Taihu Lake was sub-grade V (*lieyu wulei*), the worst in the national water quality classification system.[1] Compared with that in 2004, the water quality in all parts of the Taihu Lake did not turn any better,

[1]State Environmental Protection Administration (SEPA), Gazette of China's Environmental Situation (2005), Ministry of Environmental Protection of People's Republic of China (27, July 2006), http://www.sepa.gov.cn/plan/zkgb/05hjgb/.

Table 7.1. Taihu Lake's Water Quality Comparison between 2004 and 2005.

Lake Section	2005	2004
Wuli Lake	Sub-grade V[a]	Sub-grade V
Meiliang Lake	Sub-grade V	Sub-grade V
Western Part Near The Bank	Sub-grade V	Sub-grade V
Eastern Part Near The Bank	Sub-grade V	Grade V
Middle Area of Taihu Lake	Grade V	Grade V
Whole Taihu Lake's Average	Sub-grade V	Sub-grade V

[a] China's lake water quality is classified from grade I to grade V indicating good, fairly good, slightly polluted, poor, and hazardous. Sub-grade V is the worst in the national water classification system.
Source: State Environmental Protection Administration (SEPA), Gazette of China's Environmental Situation (2005), http://www.sepa.gov.cn/plan/zkgb/05hjgb/.

with most parts of Taihu Lake even dropping from grade V to sub-grade V (Table 7.1).

Actually, the Chinese government, including both central and the local departments, has for years paid intensive attention to pollution control in the Taihu Lake, surrounded by one of China's most affluent areas with the highest population density. Unlike the historically-untamed Huaihe River that flows across four provinces, the Taihu Lake, with better ecological basis, is mainly the internal affair of Jiangsu Province and should be easier to govern in terms of environmental protection. The Chinese government in its 11th Five-Year Plan reiterated higher pollution control standards and goals for Taihu Lake than those for Huaihe River, only to find the environment in Taihu exacerbating at a faster speed and even causing drinking water crises.

CRAZY FOR GROSS DOMESTIC PRODUCT (GDP)

Since China deepened its reform and opening-up through introducing the expression "socialist market economy" and thus unleashing market forces in early 1990s, its gross domestic product (GDP) especially in coastal areas kept a high-speed growth momentum for about 15 years, which imposed huge pressures upon the country's already worsened environment and scant resources. The Taihu Lake is surrounded by five populous cities, namely Wuxi, Suzhou, Changzhou in Jiangsu Province and Jiaxing, Huzhou in

Zhejiang Province, which jointly contribute 13% of China's GDP and 19% of the fiscal revenue.

Jiangsu's economy is ranked the third among all provincial areas on the Mainland China in 2006, only after Guangdong and Shandong Province. Jiangsu, famous for its light industry in history, started to boost its heavy industry during 1980s. In 2000, the added value of heavy industry accounted for 56.7% of Jiangsu's total industrial added value, while in 2004, this proportion rose to 65.8%.[2] The rapid development of Jiangsu's industry especially heavy industry led to rocketing increase in energy and resource consumption as well as gradual deterioration of local natural environment. In 2004, Jiangsu's industrial sectors discharged waste gas of 1.78 trillion cubic meters and wastewater of 2.635 billion tons, soaring 96.3% and 30.5% from 2000.[3] The pollution extent of surface water in Jiangsu is much higher than the average level across the country. About two-thirds of the river sections in the province witness their water quality drop below grade III in the national water classification system,[4] and clean water resources have become rarer and rarer in Jiangsu. In 2002, altogether 4.32 billion tons of wastewater were discharged in the Jiangsu Province, containing chemical oxygen demand (COD) of 784,000 tons, while in 2006, these two figures rose 19.4% and 18.7% to 5.16 billion tons and 930,300 tons, respectively.[5]

Wuxi, whose GDP and industrial output ranked the 9th and the 7th among all Chinese cities in 2006,[6] shifted its industrial focus from

[2]Jiangsu Provincial Bureau of Statistics, "The Practice and Approaches of Jiangsu Changing Its Way of Industrial and Economic Growth", (*jiangsusheng zhuanbian gongye-jingji zengzhangfangshi de shijian yu lujing*), National Bureau of Statistics (13, February 2006), http://www.stats.gov.cn/was40/gjtjj_detail.jsp?channelid=33728&record=88.

[3]*Ibid.*

[4]*Ibid.*

[5]Statistics are respectively from Jiangsu Environmental Protection Office, "Gazette of Jiangsu's Environmental Situation (2002)" (*jiangsusheng huanjing zhuangkuang gongbao 2002*), Jiangsu Environmental Protection Office (28, October 2006), http://www.jshb.gov.cn/jshb/hjzk/showinfo.aspx?infoid=b4c2fd84-ebb0-4947-a5ce-9119c8d5e4d2&categoryNum=003007&siteid=1 and "Gazette of Jiangsu's Environmental Situation (2006)" (*jiangsusheng huanjing zhuangkuang gongbao 2006*), Jiangsu Environmental Protection Office (4, June 2007), http://www.jshb.gov.cn/jshb/hjzk/showinfo.aspx?infoid=d02d65bd-0075-4669-83bd-18463b711c6c&categoryNum=003007&siteid=1

[6]Huang Xiaowei, "Wuxi, a Battle between a City and Pollution", *Southern Weekly* (2, August 2007), p. C13.

light industry to heavy industry very quickly over the past two decades. In 1980, the heavy industry only accounted for 42% in Wuxi's total industrial output, while nowadays this ratio rose to around 74%.[7] Wuxi actually follows the traditional way for industrialization, importing energy, raw materials and manufacturing chemical, steel, cement, and electroplated products that consume large amounts of energy while producing severe environmental pollution. In Yixing, a nearby small city that is affiliated to Wuxi, there are 1200 chemical plants.

Besides industrial pollution, urban sewage and agricultural pollution are also culprits for Taihu's water crisis. A significant proportion of phosphorous content came from urban sewage while nitrogen materials were from the usage of fertilizers. The urbanization process in Wuxi speeded up since 1980s, with registered urban population jumping from 810,000 in 1982 to 1.01 million in 1990 and 2.13 million in 2001.[8] Meanwhile, the floating population in the urban and rural areas of Wuxi quadrupled from 334,405 in 1993 to 1,347,274 in 2003.[9] In 1979, the fertilizer averagely used in per hectare arable land in Taihu drainage basin was 24.4 kilograms, while nowadays the figure rose to 66.7 kilograms, 70% of which flowed away through rain waters.[10]

The long-time fruitless effort in curbing Taihu Lake's pollution is only one example showing China's inefficiency in solving local ecological problems. Among various causes that may cover a wide range of realms, the most fundamental one may lie in the multilayered and multisectioned governmental structures in the country. In a state-centric country like China, the government rather than the market plays a dominant role in environmental protection; so, if the governmental institutions have innate flaws, the problem will probably become incurable.

As one of the most affluent provinces in China, Jiangsu does not lack sufficient fund or technology for environmental protection. The Taihu

[7] *Ibid.*

[8] Shanghai Population and Development Research Center, "Research Report on Wuxi's Population Development Strategy" (*wuxishi renkou fazhan zhanlue yanjiu zongbaogao*), Shanghai Population and Development Research Center, http://www.spic.sh.cn/zy_rkyj_02.htm.

[9] *Ibid.*

[10] Huang Xiaowei, "Wuxi, a Battle between a City and Pollution", *Southern Weekly* (2, August 2007), p. C13.

crisis, however, showed that its environmental governance was no better than other parts of China. Officials at all levels have realized the seriousness of Taihu problem for a long time, but the local administration's unitary goal on GDP growth greatly weakened clean-up motivations. It is the EPB/EPO's job to supervise the environmental causes, but their heavy dependence upon local government and disadvantaged positions in interaction with other departments made green targets very difficult to achieve.

Chemical factories around the Taihu Lake are the No. 1 polluters that later caused the blue-green algae outbreaks, but it is also the chemical industry that boosted the economy of Wuxi forcefully in the past two decades. There were about 20,000 chemical plants in the Taihu Lake basin before the water crisis. Wuxi is well known in China as the cradle of Township and Village Enterprises (TVEs), and many local TVEs put their focus on chemical production, making great contributions to the local economy growth. In 1996, there were 128 chemical plants out of all 184 manufacturing enterprises in the Wuxi's Zhoutie Town, nicknamed as the "town of chemical industry". The output of these chemical plants accounted for 85% of the town's total at that time. After years of closing down high-polluting enterprises, there were still over 100 chemical plants before the Taihu algae crisis, whose output was about 40% of the town's total.[11] Chen Qingfeng, party committee member in charge of propaganda in Zhoutie Town said that the stricter environmental standards in the town had caused transfer of investment worth 1 billion yuan (US$ 133 million) from Zhoutie to other places. "These 1 billion yuan would have yielded 4 billion yuan of annual sales income if kept in Zhoutie, and our total annual sales income for the town was only 6 billion yuan," Chen said.[12]

Immediately after Wuxi municipal government made the decision to close those 100 chemical plants in Zhoutie after the water crisis, agents from other regions scurried to Zhoutie, inviting those chemical enterprises to be relocated in their places with favorable conditions concerning taxation and land use. Therefore, without a unified and universally-reinforced environmental standard, the pollution can only be transferred, instead of being solved fundamentally. Statistics show if the national standard is strictly

[11]Gao Xuejun, "Taihu's Water Crisis Causes Close-Down of Chemical Plants in Wuxi", *China Business News* (5, July 2007).
[12]*Ibid.*

followed, a chemical project with an annual sales of 5 million yuan needs to invest about 1 million yuan in infrastructure construction for wastewater treatment, often exceeding the initial investment of the project. Facing huge pressures from the central government and the public, local officials had to take high-profile measures including closing down polluting factories and punishing lower-level officials, but their anxiety about the possible slowdown of economic growth and huge clean-up cost is still there. After sometime when the public attention gets shifted, the environmental standards and law enforcement will become loose again. This is the vicious circle that keeps going on and prevents Taihu Lake from turning limpid again. Stricter local environmental requirement could only drive away entrepreneurs and investors if other regions of China remain the same.

WU LIHONG'S TRAGEDY: FAILED CRUSADE AGAINST POLLUTION

Long before the carpet of algae scum on the Taihu Lake elicited national outrage in 2007, many residents living nearby the lake had realized the seriousness of local industrial pollution. Wu Lihong, an indigenous environmental activist in Zhoutie Town, started his 16-year one-man crusade against pollution in the once-beautiful Taihu Lake in early 1990s. Wu, a former sound-proofing equipment salesman who made insistent effort to pressure polluting factories and bureaucrats to clean up the lake water, was once awarded by Beijing the title of "Eco-warrior" but finally found himself in jail when the 2007 Taihu algae crisis outbreak justified his decade-long action and made local officials more embarrassed.

Mr Wu, born in 1968, grew up beside the Taihu Lake and saw with his own eyes how it had been turned from a scenic spot into a foul and poisonous lake from which people cannot drink its water. He still remembered swimming and catching water fowl in the lake when he was a boy, claiming that the chemical factories and local government officials should be blamed for the pollution over the past 20 years.[13] Wu had been documenting the dramatic increase in the pollution through taking photos

---

[13]Minnie Chan, "Wu Lihong in Front of the Environmental Protection Agency", *South China Morning Post* (29, May 2006).

and writing report. He bypassed the local bureaucracy and directly filed reports to higher-level governmental officials or the mass media, which irritated officials and factory owners with vested interest. Since the mid-1990s, more than 200 polluting factories have been closed, thanks in part to Mr Wu's efforts.[14] In November 2005, Wu was awarded as one of China's top 10 environmentalists by a panel of judges from the National People's Congress and the Chinese People's Political Consultative Conference (CPPCC) and was honored at a ceremony held at the Great Hall of the People in Beijing.

Wu himself and his family, however, had to pay dearly for such spiritual reward. Wu was beaten three times by thugs in 2003 and had two ribs broken, while his wife lost her laboratory technician's job when the chemical factory she worked in closed down in 1998 because of his report. Over the years, Mr Wu has spent more than 400,000 yuan (US$ 50,000) of his own savings to finance his one-man crusade against pollution of the lake and he lost his job after his boss said he was warned by local officials to fire him.[15]

Officials in Yixing, a county-level city affiliated to the Wuxi Municipality, confirmed Wu as a real threat in 2006 when he was sending petitions to the State Environmental Protection Administration to thwart Yixing's application for the title of model city on environmental protection. In October 2006, Yixing successfully obtained the title of "National Model City on Environmental Protection", despite wide skepticism among its residents living near the Taihu Lake. Wu Lihong refused to give up, continuing to collect photographs and samples of the polluted water as evidences for planned administrative litigation with the aim to revoke the green award on Yixing City. Wu planned to leave for Beijing to lodge the litigation on 20, April 2007, and one week before his leaving, Wu was arrested by the local police with a "blackmail" charge.

Environmental Protection officials in Beijing and Yixing held somewhat different stances toward Wu. Environmental officials in Yixing stood on the same side of local leadership, regarding Wu as a big threat toward the city's image as an environmentally-friendly city. After Wu received his green

---

[14] *Ibid.*

[15] *Ibid.*

award in Beijing, Shen Jianjiang, director of the Yixing Environmental Protection Bureau told the media that he knew nothing about the award. "I don't know anything about Mr Wu's award because he has never talked about it. He is only interested in filing reports to officials above us. If you want me to commend him ... sorry, I can only say I will not do that," Mr Shen said.[16]

SEPA officials in Beijing, however, had a mixed emotion with Wu. On the one hand, they knew very clearly that the water quality in Taihu Lake was deteriorating quickly and a large part of Wu's petitions were reasonable. They knew environmental activists such as Wu were indispensable for the improvement of China's environmental governance, and therefore held a sympathetic stand. In 2002, Wu was detained by the local police in Zhoutie Town, and with help from Tian Weiyong, director-general of the pollution control department under SEPA, he was later released. On the other hand, they themselves were pressured by other vested interests and had to be very cautious on this sensitive issue, especially after Wu granted interviews to German and other foreign media interviews about his campaign. Wu's case is reminiscent of that of Mr Jiang Yanyong, a retired army doctor who courageously exposed a cover-up of China's 2003 SARS outbreak to western media and later got a much cooler reception at home. Jiang was denied permission by the Chinese government to travel to the United States to receive a human rights award from the New York Academy of Sciences.

In June 2007, Yixing city prosecutors charged Wu for extorting 55,000 yuan from enterprises by threatening to expose how they were polluting the environment. Zhang Lijun, vice head of the SEPA, made an immediate comment on the issue by saying that "environmental protection departments always regard environment advocates as allies."[17] Two months later, Wu was sentenced to 3 years imprisonment. The tragedy of Wu and other whistle-blowers who exposed cases of environmental or public health disaster reflects the Chinese political system's limited tolerance to dissenting voices. Wu not only posed threat to the interests of local officials and business people, but also challenged the party's national grip of the media and further undercut the public's waning confidence upon the

[16] *Ibid.*
[17] "Eco-warrior Wu Lihong Charged for Blackmail", *Xinhua News* (6, June 2008).

government for information transparency and clean governance. Although the party has verbally asked citizens to report any misbehavior by officials amid rising concern over corruption, nongovernmental organizations and individual activists still take great risks when speaking out where things are going wrong. The opaque authoritarian system and hierarchical structure without sufficient supervision from individuals, media, and NGOs on many occasions prevent environmental disasters from being found at early stages. The crackdown upon civic activities will dampen people's enthusiasm in participating in public welfare as well as foster officials' numbness and selfishness. Without effective supervision from the civic society, it is very unlikely for the Chinese government to fundamentally solve the pollution problem in the Taihu Lake.

POSSIBLE SOLUTIONS

Taihu's water crisis immediately aroused public indignation as well as central government's high attention. On 11, June 2007, the State Council held a special conference in Wuxi City to discuss the solutions to the algae breakout. Premier Wen Jiabao wrote an instruction to the conference, saying the work against water pollution in Taihu had been carried on for years, but the problem was not solved fundamentally. Upon his demand for enhancing comprehensive treatment, the Wuxi municipal government started to close many chemical plants around the lake, including a hundred of them in Zhoutie Town. Meanwhile, Wuxi planned to block all the pollution discharge tubes to the Taihu Lake in 2 years, and suspended approval on any new project that discharges nitrous and phosphorous materials to the Taihu Lake. All the townships in the Taihu Lake basin are required to set up water treatment factories that reach national standards. The Wuxi municipality also established two special teams for sudden flight inspection over different districts, and these districts are required to check one another regularly on water disposal.

It seemed that the "Zero O'clock Action" in 1998 came back again, and definitely some short-term effect would come out with Taihu water's pollution alleviated to some extent. But no campaign could last long and any attention would meet fatigue after some time. Figures have shown that in most of the time in 1990s before the "Zero O'clock Action", only 540 million cubic meters of industrial wastewater and 320 million cubic meters of urban sewage were discharged into the Taihu Lake, but in 2000 alone, just 2 years after the "Zero O'clock Action", the two figures rose alarmingly to 3.24 billion cubic meters and 2.09 billion cubic meters.[18] Without any fundamental institutional changes, the pollution problem could not be easily solved with one of two high-profile campaigns, as the in-depth causes the existing current economic and political systems remain intact.

It is not objective to say that local governments in China are development-driven while the central government is pro-environment. In fact, from the central level, the government's "development first" stance has not been fundamentally changed, with the SEPA playing a minor role in decision-making while the NDRC playing a larger role. It became evident when China claimed in its first national plan on climate change that "climate change is an issue involving both environment and development, but it is ultimately an issue of development." Throughout China's multi-layered institutional structure with territorial divisions at the center, province, city, county, township, and village levels, environmental protection departments are all in inferior positions in the interagency bureaucratic interactions with their budget and personnel affairs determined by other agencies.

The Chinese government has to enhance the power of its environmental protection sectors to solve the pollution problem more effectively, with its national strategy putting real equal focuses on economic development and environmental protection. It should depend on introducing economic incentives and disincentives to curb pollution and ecological destruction instead of relying too much on short-term administrative orders such as closing polluted enterprises or blocking pollution sources. Institutional innovations such as levy of reasonable discharge fees, resource, fuel, and environmental taxes can do a better job than administrative orders as they are fair and equal to every plant, and their effect can last very long. More public participation is also necessary for environmental

[18]Deng Jiansheng, "Taihu's Pollution Continues to Deteriorate Alarmingly after the Zero O'clock Action" (*lingdian xingdong hou chixu ehua, taihu wuran yiran chumujingxin*), *Huadong Xinwen* (8, December 2004), p. 8.

protection. In China's state-centric society, the forces of environmental non–governmental organizations (NGOs) are still very weak and the media's supervision is quite limited, because the government has great influences on them. The usual case in China is that many problems can only be noticed after a real crisis breaks out. Without any institutional changes, introduction of unified economic incentives and disincentives, or active public participation, no one can promise that the Taihu Lake algae crisis would not happen again.

Chapter

8

Case Study: Beijing Olympics and Air Quality

Beijing hosted the 2008 Summer Olympics, officially known as the Games of the XXIX Olympiad, from 8 to 24, August 2008. As an event of paramount importance to socialist China, the 2008 Olympic Games with related spending of around US$ 35 billion was regarded by Beijing as a showcase of its rapidly-growing economic strength and rising international status, as well as an indicator of the communist regime's legitimacy.

The Chinese government took great steps to ensure the complete success of the pageant, but the capital's poor air quality that all athletes and attendees had to face in the summer had been an area of concern for the organizers and for China's top leaders.

SMOG CASTS SHADOW OVER "GREEN OLYMPICS"

China is one of the world's fastest-growing economies, but its cities pay for it with choking pollution. Beijing is often blanketed by gray smog, some from coal-fired industries and construction sites, and others from neighboring provinces or rocketing vehicle numbers.

Table 8.1. Beijing's Major Air Pollutant Concentration (Unit: milligram/cubic meter).

	Year 2005	Year 2006	% Change (2006/2005)
Small particulate matter (PM)	0.142	0.161	13.4%
SO_2	0.050	0.053	6%
CO	2.0	2.1	5%
NO_2	0.066	0.066	0%

Source: Statistics for Year 2005 and Year 2006 are respectively from "Beijing's Environmental Statement (2005)", Beijing Environmental Protection Bureau (26, September 2006), http://www.bjepb.gov.cn/bjhb/tabid/68/InfoID/7622/frtid/375/Default.aspx and "Beijing's Environmental Statement (2006)", Beijing Environmental Protection Bureau, May 2007, http://www.bjepb.gov.cn/bjhb/portals/0/fujian/zwgk/2006.pdf.

Small particulate matter (PM) — fine particles of pollution that can easily pass into people's lungs and harm health — is the major pollutant in Beijing's atmosphere and a main cause of the capital's dark gray sky. A UN report found that the average level of small PM in Beijing's air in 2006 was eight times higher than the level recommended by the World Health Organization.[1]

Other forms of air pollution, including sulfur dioxide, nitrogen dioxide, and carbon monoxide (Table 8.1), registered increases in 2006 following 3 years of decline, according to official statistics provided by the Beijing Environmental Protection Bureau (EPB) in its annual gazettes.[2] Fine PM concentrations in Beijing also grew 13.4% year on year in 2006.

Smog in Beijing is not merely an image problem, but one that will affect the normal performances of athletes during the Games. Some athletes who competed in the Olympic test events in 2007 complained of foul air leading to respiratory difficulties and causing upper-respiratory infections and nausea.[3]

[1]"Foul Beijing Air may Delay Some Olympic events", *AP News* (25, October 2007).

[2]Beijing and other provincial-level areas are required by China's Environmental Protection Law to release their environmental statements annually. Beijing Environmental Statements (*beijingshi huanjing zhuangkuang gongbao*) are available on the official website of the Beijing EPB, http://www.bjepb.gov.cn/bjhb/portal0/default375.htm.

[3]Juliet Macur, "2008 Olympic Challenge: Breathing", *International Herald Tribune* (25, January 2008), p. 4.

Colby Pearce, 35, an Olympic hopeful in track cycling from Boulder, Colorado, said in 2007, he saw smog floating inside the velodrome in Beijing, and his throat became scratchy and he developed bronchitis because of air pollution.[4] A US boxing team, while competing in China in December 2007, ran in the hotel hallways instead of on the streets due to serious outdoor air pollution.

An athlete's reaction to pollution exposure is immediate. With bad air, the defensive mechanism of the body will make the athlete breathe less of it, leading to oxygen debt and even disastrous consequences during the contests. Serious pollution can also provoke allergic reactions, or trigger an asthma attack, and the risk of a heart attack increases on high-pollution days. An athlete working out at a moderate pace for 30 minutes in poor air is equivalent to a sedentary person breathing that same air for 8 hours.

When "particulate matter" is breathed in, the particles deposit on the lungs and can pass through the lungs into the bloodstream. August is one of the hottest and most humid months in Beijing; the over 30°C heat, 95% relative humidity, together with the smog in a windless season will be extremely hazardous for endurance athletes — marathoners, triathletes, and cyclists — who have to compete outdoors for hours.

In a visit to China in August 2007, Jacques Rogge, the president of the International Olympic Committee, said some of the Beijing Olympic events might have to be postponed if the athletes are at risk.[5] Rescheduling those events, particularly at the last minute, could be a disaster for athletes and coaches.

Some sports federations, like those of the United States, Australia, and Britain, had announced that some of their athletes would not arrive in Beijing until right before the competitions to avoid being exposed to poor air quality. Some Olympic teams were scouting for possible alternative training sites in South Korea, Singapore, Japan, and Malaysia for use in the days before the Beijing Games.

---

[4] *Ibid*.

[5] Juliet Macur, "Beijing Air Raises Questions for Olympics", *The New York Times* (26, August 2007).

Growing international concerns over Beijing's air pollution embarrassed the organizers of the Games, imposing a huge challenge toward the "Green Olympics" concept put forward by Beijing years ago.[6]

QUESTIONABLE "BLUE SKY" ACHIEVEMENTS

At the beginning of 2008, the Chinese government rolled out new statistics claiming that air quality had dramatically improved since its successful Olympics bid in 2001, and that more "blue sky" days can be seen in Beijing now. According to the standard set by the State Environmental Protection Administration (SEPA), a "blue sky" day in Beijing is when the Air Pollution Index (API) (Table 8.2) is equal to, or lower than 100.

Table 8.2. China's Air Pollution Index System.

	Pollutant Concentrations (milligram/cubic meter)				
Air Pollution Index	SO_2	NO_2	PM10[a]	CO	O_3
50	0.050	0.080	0.050	5	0.120
100	0.150	0.120	0.150	10	0.200
200	0.800	0.280	0.350	60	0.400
300	1.600	0.565	0.420	90	0.800
400	2.100	0.750	0.500	120	1.000
500	2.620	0.940	0.600	150	1.200

Source: "China's Air Pollution Index Definitions and Limits", Ministry of Environmental Protection of People's Republic of China, http://www.sepa.gov.cn/quality/background.php.
Note: Various pollutants' API will be calculated individually and the largest among all the five APIs will be regarded as the city's final API.
[a]Fine Particulate Matter (PM), considered as either PM2.5 (particles with average diameters ≤ 2.5 micrometers or PM10 (diameters ≤ 10 micrometers), is directly emitted from power plants, motor vehicles, industrial facilities, and other sources.

[6]Beijing has put forward three concepts for the Beijing Olympic Games: Green Olympics, High-tech Olympics, and People's Olympics. According to the Beijing Organizing Committee for the Games of the XXIX Olympiad (BOCOG), "Green Olympics" means that priority will be given to the environmental protection (EP) in the planning, designing, and construction of Olympic venues and facilities, and that strict ecological environment standards and systematic supporting systems will be established. EP technologies and means will be extensively used in promoting, on a large scale and in various aspects, environmental treatment, afforestation, and beautification of urban and rural areas.

The Chinese government has been reporting daily air pollution levels of major cities for years, on the basis of the data collected by SEPA and its local branches from monitoring stations around those cities. Local environmental protection bureaus then calculate the API, indicating the potential harm to human health at a range of 1–500. The higher the API, the more polluted is the air.

According to the air quality statistics released by the Beijing EPB, the number of "blue sky" days (API $\leq$ 100) jumped from 185 in 2001 to 246 in 2007, an increase of 33% over 6 years (Figure 8.1). Beijing had targeted 230 "blue sky" days in 2005, 238 in 2006, and 245 in 2007. All these goals have been realized as announced by its EPB.

Foreign environmental experts have cast doubts on whether air quality has truly improved in Beijing and has concluded that methodology changes in the measurement of air pollution have enabled the city to meet "blue sky" targets linked to the Olympic Games.

The Beijing API is based on an average of data from selected monitoring stations; so, the location of those monitoring stations picked up by the statisticians will greatly influence the final result of API. APIs in high

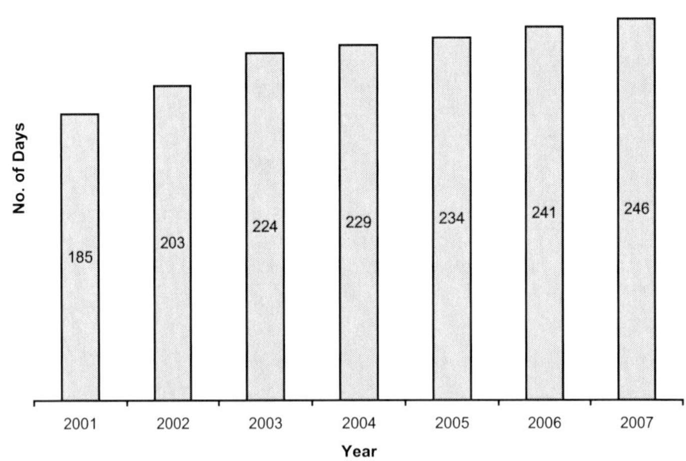

Figure 8.1. Officially-Announced Number of "Blue Sky" Days in Beijing from 2001 to 2007.

Source: Statistics for Year 2001 to Year 2007 are from Beijing's Annual Environmental Statements which are available on the official website of the Beijing EPB, http://www.bjepb.gov.cn/bjhb/portal0/default375.htm.

traffic and densely-populated areas as well as industrial districts are certainly higher than other suburban places.

According to western journalists, from 1998 to 2005, the same seven stations — located in the city center — were used to measure air quality. In 2006, however, when international scrutiny on China's air quality was increasing, two stations monitoring traffic pollution were dropped from the city API calculation, while three additional stations in less polluted areas were added.[7]

Calculating the Beijing API values for 2006 and 2007 using data from the original monitoring stations would have changed the outcome considerably; in fact 38 of Beijing's 241 so-called "blue sky" days in 2006 would not have been qualified as "blue sky" under the old methodology. The number is even less for 2007: 55 fewer days would have attained the "blue sky" standards out of the 246 reported "blue sky" days that translates into fewer "blue sky" days as a whole than in 2002 (which had 203 reported "blue sky" days) (Figure 8.2), immediately after Beijing was awarded the Olympics.[8]

The Western media reported that the Chinese government also used less stringent standards to measure air pollution in Beijing. From June 2000,

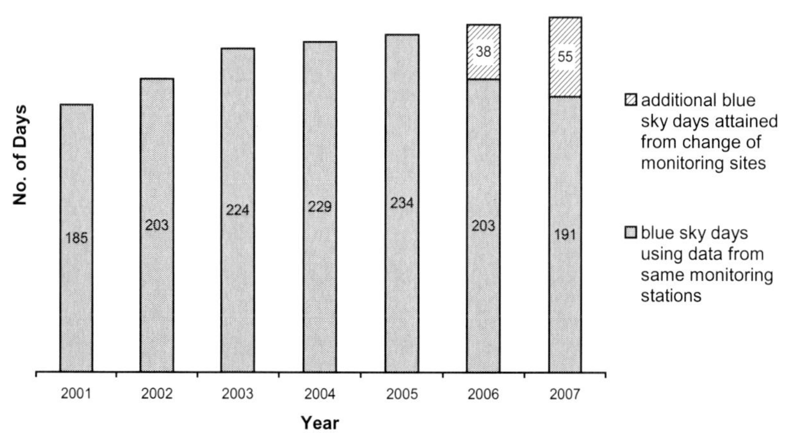

Figure 8.2. Adjusted "Blue Sky" Day Counting in Beijing.
Source: Steven Q. Andrews, "Beijing's Sky Blues", The Wall Street Journal (9, January 2008).

[7] Steven Q. Andrews, "Beijing's Sky Blues", *The Wall Street Journal* (9, January 2008).
[8] *Ibid*.

measurements of nitrogen oxides (NO_x) were replaced by the much less stringent nitrogen dioxide (NO_2) measurements.

Beijing was also counting more days just below or just above the "blue sky" boundary (API 96-105) as "blue sky" days. In 2001, Beijing had 34 days where measures for PM10 were near the blue sky boundary, and approximately half were reported as "blue sky" days.[9] In 2006, 49 days had PM10 values equivalent to an API of between 96 and 105, and 98% of those were reported as "blue sky" days.[10]

SOURCES OF AIR POLLUTION

The slow pace of improvement in Beijing's air quality can be mainly attributed to the extensive use of coal, the growing number of motor vehicles, and the city's geographical location. The smog in Beijing is just another telling example of China's environmental deterioration against the backdrop of double-digit economic miracles.

As the primary energy source, coal provides 69% of China's energy needs. About 90% of China's sulfur dioxide emissions and 50% of its particulate emissions are the result of coal use. Although it has eliminated the use of small coal-fired boilers in downtown Beijing to reduce soot and sulfur dioxide emissions, coal is still allowed to be used in industrial facilities and large heating boilers in farther places. Coal still makes up 40% of Beijing's total energy consumption even though it is less dependent upon coal than other Chinese cities.

Neighboring Hebei, Shandong and Shanxi Provinces, and the Tianjin Municipality notorious for their coal-fired heavy industries also have a large part to play in Beijing's poor air quality. Beijing is surrounded by the Hebei Province, which is responsible for contributing between 50% and 70% of Beijing's PM2.5 concentrations and 20%–30% of ozone during sustained wind flow from the south in summer.[11] A study shows that in the summer of 2001, Hebei, Shandong, Tianjin, and Shanxi contributed 32%, 11%, 13%, and 3.5%, respectively to the PM2.5 concentrations in

[9] *Ibid.*
[10] *Ibid.*
[11] David G. Street *et al.*, "Air Quality During the 2008 Beijing Olympic Games", *Atmospheric Environment*, Vol. 41 (2007), p. 480.

Beijing.[12] This means that controlling local pollutants alone in Beijing will not be sufficient to attain air quality goal set for the Beijing Olympics.

Beijing's geographic location is also a decisive factor even if it does not generate man-made emissions: levels of fine PM and ozone could still be high and could exceed healthful levels under unfavorable meteorological conditions. When wind speeds decelerate in summer, the hills surrounding Beijing will restrict the venting of pollution, exacerbating the air quality in the capital.

Another major source of Beijing's smog is the fast-growing number of vehicles on the roads and the corresponding congestion problem. Over the past 7 years, the number of vehicles in Beijing almost doubled from 1.58 million in 2000 to 3 million in May 2007 (Figure 8.3).[13]

About 40% to 50% of the major pollutants in Beijing's air — nitrogen oxides, carbon monoxide, and inhalant particulate matter — come from

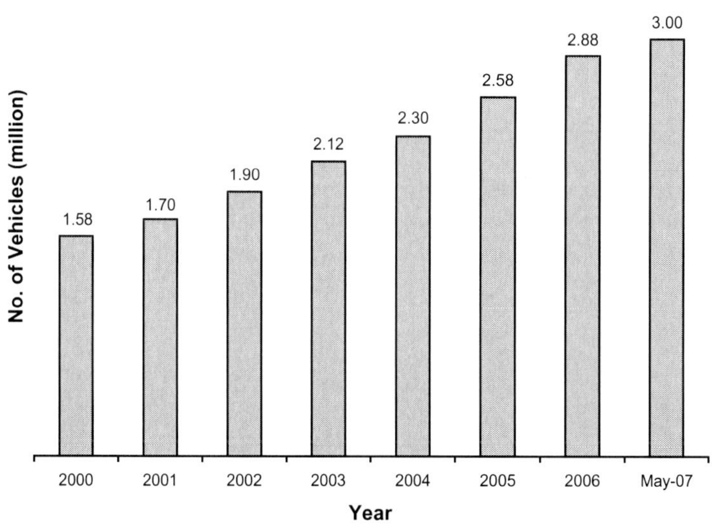

Figure 8.3. Number of Vehicles in Beijing (Unit: Million).
Source: Beijing Statistical Bureau.

[12] *Ibid*., pp. 489–490.

[13] The figures are from the Beijing Statistical Bureau.

vehicle exhaust emissions, according to Zhu Tong, member of the expert panel for Olympics air quality protection and professor of the environment science and engineering school at Beijing University.[14]

CONTINGENCY MEASURES AND SOLUTIONS

Beijing has spent US$ 15 billion battling environmental pollution, more than it promised when the city won the Olympics bid in 2001. The capital as well as the country is trying to promote public awareness on the importance of environmental protection. China, in 2007, established the National Leading Group to Address Energy Saving, Emission Cutting and Climate Change, headed by Premier Wen Jiabao.

Some leading indicators of environmental degradation in China showed signs of improvement. In the year 2007, China's two major pollutants, SO_2 and chemical oxygen demand (COD), dropped for the first time by 4.66% and 3.14%, respectively, year on year, and per unit GDP energy consumption decreased by 3.27%, according to a SEPA report.[15]

Green measures paid off to some extent, but if Beijing wanted to guarantee dustless air during the Games, some contingency plans must be worked out in advance. Official report from Beijing Municipal Environmental Protection Bureau showed that in May 2008, the city's air quality was substandard for 18 days of the month, an increase of 7 days year on year.[16]

Limiting emissions from vehicles was the top priority of the city's pollution-curbing efforts. As early as 1 year prior to the Games, Beijing had been planning to remove half of the city's 3.3 million vehicles during the Games to improve air quality and ease traffic flow.[17] On 20, June 2008, less than 50 days to the Games, Beijing authorities officially announced that

---

[14]"Beijing to Introduce Tougher Auto Emission Standards Ahead of Olympic Games", *Xinhua News* (8, November 2007).

[15]"Premier: China Cuts Energy Consumption, Pollution in 2007", *Xinhua News* (5, March 2008).

[16]Beijing EPB, *Monthly Report on Beijing's Air Quality* (May 2008), http://www.bjepb.gov.cn.

[17]Jim Yardley, "Beijing Plans to Halve Traffic During Games", *International Herald Tribune* (23, January 2008).

starting 20 July, cars would be allowed on the roads on alternating days, based on even-and-odd-number license plates.[18] The system was in effect until 20 September, almost a month after the close of the Olympics and 3 days after the closing ceremony of the Paralympics.

High-emission or "yellow-labeled" vehicles, most of which were freight trucks, were banned from Beijing roads from 1 July to 20 September to ensure a green Olympics and Paralympics as promised by the authorities. Yellow labels are displayed on the windscreens of more than 300,000 vehicles in Beijing that fall short of the Euro I emission standard, the lowest of the European emission standards.

Temporary traffic restrictions had been anticipated for the Olympics since August 2007, when Beijing conducted a four-day experiment that limited motorists to driving on alternate days, depending on whether their license plates end in an odd or even number.

That test eliminated more than one million vehicles each day, easing traffic but having a less dramatic effect on air pollution. Although the Beijing municipality insisted that the air quality during the four-day odd–even test was better than normal according to figures from monitoring stations, many local residents claimed that the air in those days was not anything special. If this kind of car restriction policy could have been extended to a longer period, e.g. 3 months, it may have a more substantial impact upon Beijing's air quality.

The city had dedicated special lanes to Olympic traffic and increased public transportation with new shuttle buses to accommodate visitors and local residents. Encouraged by the government, many civil servants in the capital took leave to reduce commuting traffic during the Games.

Special measures enacted by the government also included the shutdowns of factories in the outskirts of Beijing and neighboring provinces in northern China during the Games to reduce spillover effect of air pollution. To reduce dust and fog from construction sites, Beijing has stopped almost all building works except those related to Olympics since the beginning of 2008.

---

[18]Mei Fong, "China Maps Out New Route to Cleaner Air", *The Wall Street Journal* (23, June 2008).

Amid recent global upsurge of climate-related anxiety, Chinese leaders planned to use the Beijing Olympic Games as a showcase for the country's devotion to green causes. The first "carbon neutral" summer games involved the extensive use of solar power, planting of trees, banning of cars from the streets, and reducing emissions from enterprises. The games, said officials, would produce 1.18 million tons of CO_2 and the counter-measures would reduce 1.03–1.30 million tons.[19] During the Games, Beijing's key environmental achievements included the extensive use of renewable energy such as solar, geothermal, and wind power in the Olympic venues, adoption of stricter vehicle emissions standards, banning of cars, development of public transportation, afforestation and water, sewage disposal.

About 20% of the Olympic venue electricity used during the Games came from the 64,500-kilowatt Guanting wind power station, Beijing's first wind power plant that could help cut yearly emission of CO_2 by 100,000 tonnes and save 50,000 tonnes of coal each year.[20] Beijing increased its new vehicle emissions standards to Euro IV ahead of schedule for the Games in an attempt to improve air quality, which was amongst the most stringent emissions standards in the world. The city opened four new subway lines and a direct line within the Olympic Green to encourage public transportation.

Along with other low-energy vehicles, Beijing also launched a fleet of 3759 buses running on Compressed Natural Gas.[21] This is one of the largest fleets of this kind operating anywhere in the world. By the end of 2007, 16,000 boilers under 20 tonnes and 44,000 boilers under one tonne had been upgraded.[22] Meanwhile, 32,000 household heating systems had been converted from household coal heating to electricity.[23]

Many Olympic sponsors, such as Coca-Cola, Panasonic, Samsung, Lenovo, and Haier, also took this opportunity to showcase their leadership in promoting environmentally-friendly products for the Games. During the Games, Coca-Cola, Haier, and Sanyo showed their commitment toward

---

[19] "China, India and Climate Change", *The Economist* (7, June 2008), p. 30.

[20] "20% of Power Supply for Olympic Venues to be Wind-Generated", *Xinhua News* (20, July 2008).

[21] Greenpeace Report (2008): "China After the Olympics: Lessons from Beijing", p. 4, http://www.greenpeace.org/raw/content/china/en/press/reports/green.pdf.

[22] *Ibid.*

[23] *Ibid.*

complete phase out of the use of hydrofluorocarbon (HFC) as refrigerant which, although nonozone depleting, incorporates a high global warming potential (GWP). China-based Haier provided Coca-Cola with HFC-free coolers, which could be seen throughout Olympic venues. Such a move imposed pressure upon sponsors to the 2012 London Olympics, which is expected by many environmentalists to be the first HFC-free Games in history. During the Games, the United States awarded the Beijing Olympic Village its own gold medal for its environmentally-friendly design and efficient use of energy. US Treasury Secretary Henry Paulson presented the Leadership in Energy and Environmental Design (LEED) Gold Award to Chen Zhili, mayor of the Olympic Village and a vice-chairwoman of the National People's Congress, calling the village "an example of US–China cooperation on energy and environmental issues."[24]

Before the Games, many of Beijing's biggest polluting factories had been moved out of the capital, including China's leading steel manufacturer Shougang Group, which began relocating its facilities to north China's Hebei Province in 2005. All current Beijing-based steel operations were moving to Hebei Province, with only the headquarters, research and development facilities, sales departments, and the logistics center remaining in the capital.

Artificial rain before the start of the Games was on the agenda of the organizers to clean up air and ensure a sunny Olympics. In May 2006, Beijing boasted having generated rainfall to clear the air and streets following the worst dust storm in a decade. Technicians in Beijing usually fire several rocket shells containing cigarette-size sticks of silver iodide over the city's skies to provoke chemical reaction in the clouds and force rains.

"When conditions permit, we will artificially increase rainfall" in the days before the Olympics through a process known as cloud-seeding, said Wang Yubin, an engineer with the Beijing Meteorological Bureau.[25] "Rainfall is a way to naturally clean the air".

Despite such final push for clean air, noticeable change did not arrive in the count-down period. Beijing's Olympic village, built to house 16,000 athletes from around the world, was officially opened under a cloak

[24]"Green Olympic Village wins US gold medal", *China Daily* (14, August 2008).

[25]"China to Force Rain Ahead of Olympics", *The AP News* (25, April 2007).

of smog on 27, July 2008. In the first 10 days (20–29 July) after Beijing adopted the even-and-odd-number traffic control, its officially-announced API was respectively 55, 65, 67, 85, 113, 110, 118, 113, 96, and 90. From 24 to 27 July, the API had stayed above 100, the national standard for good air quality. Local environmental officials attributed the four-day poor air quality mainly to the hot and sultry weather with occasional breeze. To explain why Beijing had not experienced many "blue-sky days" before the Games, Du Shaozhong, deputy director of the Beijing Environmental Protection Buearu, told reporters at the Main Press Center for the Games on 26 July that "good air quality does not necessarily mean blue sky, and we should judge whether there is pollution by scientific statistics, not by what eyes could see". "It's just like you can't see things in a balneary, but you would not believe there is pollution."[26] "Technically, the haze was partly caused by rainy weather", Guo Hu, head of the Beijing Municipal Observatory, said at the same press conference.[27] Du Shaozhong discouraged people from wearing face masks during the Games, saying it was "totally unnecessary".

While verbally assuring the public, the Chinese government had to prepare the last-ditch contingency measures in case of the worst situation. On 31, July 2008, the Ministry of Environmental Protection and governments of Beijing, Tianjin, and Hebei Province jointly issued a public announcement on emergency pollution control measures including shutting another 200-plus factories and further restricting vehicles across Beijing, Tianjin city, and surrounding Hebei Province. In Beijing, besides current restrictions banning odd/even license plates on alternate days, automobiles whose last digit matches the last digit of the date would be banned. Tianjin and Hebei would begin implementing similar odd/even restrictions.[28] The announcement said that once the Games began on 8 August, the contingency measures would kick in if the authorities decide that the air quality had not improved enough.

These contingency measures definitely had some effect on mitigating Beijing's air pollution during the Games. Though foreign athletes and

[26] "Beijing Promises Air Quality During Games", *Xinhua News* (26, July 2008).
[27] *Ibid.*
[28] "China Plans Emergency Measures for Pollution", *AP News* (31, July 2008).

tourists still had worries about the air problem in August, the short-term air quality in Beijing was partly controllable considering the Chinese government's strong role in intervening domestic economic and social activities.

According to the official report released by the Beijing Municipal Environmental Protection Bureau, the capital had seen its cleanest air in a decade during the 17 days of the Games, all of which met Chinese National level II Air Quality Standard, or the "blue sky" standard that poses no harm to human health.[29] The Beijing Municipality claimed that in August 2008, the major pollutant density in the atmosphere on average dropped 45% year on year, with only 1 day's air quality not meeting the national standard.[30] During the Games, the density of sulfur dioxide, carbon monoxide, and nitrous dioxide all reached the standards of the cities in developed countries.[31]

It is a pity that despite a series of long-term measures including industrial facility improvement and moving out steel mills, Beijing finally had to resort to strict temporary measures such as traffic control to clean up air during the Games. It is not likely that Beijing's air will maintain the same quality after those temporary regulations are scrapped. Despite the officially-announced green achievement made during the Beijing Olympics, it is not realistic to expect the 17-day Games to improve Beijing's air quality or even China's environmental performances fundamentally. Rapid industrialization and urbanization as well as flawed governance system are still major causes of deteriorating environment. More effort is needed to prevent people's green enthusiasm during the Games from becoming a-flash-in-the-pan. Air quality of Beijing as well as many other Chinese cities is still not within the healthy range, according to the standards of the World Health Organization. Due to the limited transparency and a lack of independently verified data, air quality reports provided by the Beijing Municipality are problematic and western environmentalists

---

[29]Beijing Environmental Protection Bureau's Air Quality Report (1, September 2008), http://www.bjepb.gov.cn/bjhb/tabid/68/InfoID/17391/frtid/40/Default.aspx.

[30]*Ibid.*

[31]*Ibid.*

have suspected that Beijing changed methodology and sites of monitoring stations to get better air data. If Beijing wants to fundamentally solve the problem of air pollution, it needs both patience and courage to gradually change the flawed environmental governance systems at the national level.

Chapter

9

China's Climate Change Politics

Climate Change, mostly caused by increased greenhouse gas (GHG) emission from human activities, is one of the most serious environmental problems confronted by the international community today. Scientists believe that continuous temperature rise will lead to frequent meteorological disasters including: drought, flood, and tropical storms — rising sea levels that may submerge coastal cities and island states and even transform entire ecological systems, and nurture the insect population — leading to epidemic outbreaks. The topic had become quite heated in 2007, the 10th anniversary of the birth of the Kyoto Protocol that set mandatory targets for industrialized nations to cut emissions, as new reports released by the Intergovernmental Panel on Climate Change (IPCC) in 2007 described the connection between climate change and human activity with a high degree of certainty.

China has been under great scrutiny in the international arena, more so than the other industrialized nations. Estimated by the International Energy Agency (IEA) to become the world's largest emitter of greenhouse gases by the end of 2007, the world's most populous nation has been identified as one of earth's most crucial offender. China estimated its GHG emissions reached 6.1 billion tons of CO_2, equivalent to that in 2004,[1] accounting for 12.45% of the world's total 49 billion tons.

Climate change can be defined as an international political affair between nations, for which the remedial strategy is to be found in the domain of global politics.[2] As the topic of climate change has become a top concern for the G-8, APEC, and other important diplomatic fronts, it is timely and necessary to study China's diplomatic strategy on climate change and factors that influence its policy-making process.

CHINA'S PERFORMANCE IN PRE-KYOTO CLIMATE NEGOTIATIONS

Start of International Climate Talks under the UN Framework

To prevent possible climate catastrophes, the international community started negotiations under the framework of the United Nations to draft international treaties restraining GHG emissions since early 1990s. A milestone in this process was the signing of the UN Framework Convention on Climate Change (UNFCCC) at the 1992 Earth Summit in Rio de Janeiro.

The UNFCCC, with the anticipation of subsequent agreements establishing more concrete obligations, aims to achieve "stabilization of greenhouse gas concentrations in the atmosphere at a level that would prevent dangerous anthropogenic interference with the climate system."[3] The parties to the UNFCCC can be roughly categorized into two groups: Annex I members of industrialized countries and non-Annex I developing countries. The UNFCCC set basic guidelines for future international climate negotiations, but heated quarrels and sharp divergences still

---

[1]National Development and Reform Commission (NDRC), *China's National Climate Change Program*, China Climate Change Info-Net, June 2007, p. 6, http://www.ccchina.gov.cn/WebSite/CCChina/UpFile/File188.pdf.

[2]Susan Martens, *Public Participation with Chinese Characteristics: Citizen Consumers in China's Environmental Management*, in Neil T. Carter and Arthur P.J. Mol (eds.), Environmental Governance in China (London: Routledge, 2007), p. 68.

[3]United Nations, *United Nations Framework Convention on Climate Change (1992)*, United Nations Framework Convention on Climate Change, p. 4, http://unfccc.int/resource/docs/convkp/conveng.pdf.

remained among different parties throughout the international climate negotiations due to the fear that emission cutting will hamper national economic growth and incur huge cost.

Appropriate negotiation strategy is extremely important for a sovereign state to fulfill its national interest in the international climate negotiations: on the one hand, whether a country can adopt a successful negotiation tactic to attract others to its side and gain the maximized diplomatic support are essential for its final gains from the marathon-like multilateral talks, and on the other hand, to what extent the negotiation outcome, coupled with new international institutions, can affect the disposition of a state, i.e. enhancement of its soft power in the eyes of other states.

China's Negotiation Stance

China took part in the UNFCCC negotiations from its preliminary stages, viewing the climate change as a controversial issue that involved North–South equity problems. Historically, China believed that developed countries should take responsibility for environmental protection and allow developing countries the opportunity to advance as much as the developed countries.[4] As the world's largest developing country and a permanent member on the UN Security Council, China has been playing a leading role among developing countries in steering international climate change negotiations, uniting with other developing countries to form the "bloc of G-77 (the Group of 77 developing countries) plus China," which comprises about 90% of the developing countries to the FCCC.

When the Intergovernmental Negotiating Committee (INC) meetings in preparation for the UNFCCC began, China argued for equity on historical grounds, placing the blame primarily on the North.[5] Since June 1991, China aligned itself closely with India; pushed for a fair convention arguing that the South had much lesser responsibility than the North, whose high per capita emissions and opulent lifestyles of the

[4] Yuka Kobayashi, "Navigating between 'luxury' and 'survival' emissions", in Paul G. Harris (ed.), *Global Warming and East Asia* (London: Routledge, 2003), p. 88.

[5] C. Dasgupta, "The climate change negotiations in negotiating climate change: the inside story of the Rio Convention", in I. Mintzer and J. Leonard (eds.) (Cambridge: Cambridge University Press, 1994), p. 133.

people were the primary causes for climate change. China strongly opposed legally binding emission reduction targets for developing countries and insisted on industrialized countries' technology and fund aid to developing countries. It advocated the preservation of national sovereignty, claiming that developing countries have the right to develop and should neither be obliged to undertake measures that might impede development nor be subject to conditions being placed on aid or development financing.

As early as the 1992 Rio Earth Summit, China and India, speaking on behalf of other developing countries, insisted that developed countries should commit themselves to technology transfer as a requirement for developing countries' support for the proposed agreements on global warming. These stances are consistent with those of other developing countries and are embodied in the UNFCCC. China's adherence to the principle of "common but differentiated responsibilities" won the support and praise from developing countries, asserting its legitimacy and enhancing its soft power. By consolidating a coalition, China could assure that its interests would be better supported, and that it would gain international prestige in being proactive in environmental protection.[6]

Meanwhile, China also adjusted its stance when negotiating with developed countries, showing its respect of international norms of environmental protection and trying to reduce western countries' hostility toward the country after the Tiananmen incident in 1989. China felt diplomatically isolated in the early 1990s. The climate change issue came into focus at the same time, thereby serving as an ideal tool for China to regain its international position.[7] The UNFCCC negotiations were high-profile negotiations, putting China in the spotlight whenever it took the initiative.[8] China was not only one of the first countries to formulate a domestic "Agenda 21" in 1994, mirroring the UN Conference on Environment and Development (UNCED) program on sustainable development, but also passed the Air Pollution Law in 1996, accounting for provisions in the UNFCCC.[9] China

[6] Yuka Kobayashi, "Navigating between 'luxury' and 'survival' emissions", in Paul G. Harris (ed.), *Global Warming and East Asia* (London: Routledge, 2003), p. 88.
[7] *Ibid.*, p. 91.
[8] *Ibid.*
[9] *Ibid.*

thought it would gain international prestige by being proactive in the environmental protection cooperation.

Though refusing initially to accept any commitments, China and other developing countries finally agreed to the UNFCCC: to establish national emissions inventories, report on national programs, promote cooperation, and form sustainable development and information exchange.[10] Due to long-time suspicion that western countries may shift their own responsibilities and use environmental issues to hamper China's economic rise, China used to oppose a new protocol on the grounds of scientific uncertainty, or the "joint implementation" programs, fearing that giving credit for reduced emissions may lessen developed countries' commitment to emissions abatement. With international talks in progress, China gradually changed its tough stance on these issues in the following Conferences of Parties (COPs) to the UNFCCC, paving the way for the 1997 Kyoto Protocol.

CHINA'S GAINS FROM KYOTO PROTOCOL

After years of heated debate, delegates from 149 countries and regions passed the Kyoto Protocol in Japan, December 1997. The Kyoto Protocol is the first international treaty to set detailed legally-binding obligations for industrialized nations to reduce their greenhouse emissions. China completed its domestic ratification process in 2002 and appealed to other reluctant parties to ratify the protocol as soon as possible. The Kyoto Protocol went into effect in 2005, requiring 38 industrialized countries to reduce their greenhouse gas emissions between 2008 and 2012 by an average of 5.2% below 1990 levels. In practice, the Kyoto Protocol milestone not only helped China to gain diplomatic ground, but also offered economic lucre under the Clean Development Mechanism (CDM) and pushed the country to improve its environmental governance domestically.

---

[10]United Nations, *United Nations Framework Convention on Climate Change (1992)*, United Nations Framework Convention on Climate Change, pp. 5–6, http://unfccc.int/resource/docs/convkp/conveng.pdf.

Diplomatic Scores

Support from Developing Countries

The principle of "common but differentiated responsibilities," advocated by China, India, and other developing countries, was followed by the Kyoto Protocol, which argued for the CDM to encourage the developed countries to provide financial and technical assistance to developing countries. Conversely, developing countries had no concrete abatement commitments. They only needed to establish national emissions inventories, report on national programs, and promote cooperation, sustainable development and information exchange.

Since the outcome of the international climate talks in 1997, most guidelines insisted by the "bloc of G-77 plus China" have been included into the Kyoto Protocol, which only imposed compulsory emission-cutting obligations upon developed countries. As the lead negotiator of the bloc, China's persistence and tactics throughout the tough talks have won admiration and support from the Third World. The Kyoto Protocol has been a success for China's foreign policy, showing that the country was capable to use diplomatic power to formulate international rules serving its own interest. Clearly, China does not have enough hard power to coerce other developing countries to support its stance or force developed countries to compromise on the climate issue. Its negotiators had to understand the opposing viewpoints of the various camps and balance between compromise and perseverance to achieve the common goal in cooperation with the developing countries.

Praise from Some Industrialized Countries

In contrast to the United States' withdrawal from the Kyoto Protocol, China's active participation in the international cooperation on global warming has won applause from European and other industrialized countries. China's support for global cooperation to curb global warming has special meaning to the international community because of its large GHG emissions and fast-growing energy consumption. China's active participation in climate change cooperation is closely related to its concern to preserve a favorable international image as a "responsible major power." Enhancing China's international image and elevating its international stature are important goals for Chinese foreign policy. China traditionally considers solidarity with the developing countries as fundamentally important, and the climate change issue has provided an unprecedented opportunity for China to boost its prestige and shore up support from developing countries as well as enhance its relationship with the developed countries.[11] China's gradual change from suspicion to support of the climate change treaty helped to improve its relations with European countries in the early 1990s, given the ever present residual hostilities affecting such ties stemming from the Tiananmen incident.

Soft Power Enhancement through Institutional Participation

China's ratification of the Kyoto Protocol shows that it is willing to accept the international norm of environmental protection, which has been gradually developed in the international community since the 1950s. China's support for the Kyoto Protocol, one of the most important international institutions in the environmental area, boosted its soft power in the international arena.

Without new international institutions that provide incentives and disincentives for individual states, sovereign states tend to adopt free-ride strategy and refuse to participate in joint efforts against global warming when the domestic emission-cutting cost is too high. International institutions such as the Kyoto Protocol, therefore, become essential to motivate individual states to collective actions for global welfare and overcoming the free-ride preference. Refusing to join international cooperation and institutions against such negative externalities will be regarded by other states as selfish and irresponsible. The country's reputation in the international community will definitely be weakened if it refuses to shoulder its due responsibility on global public issues.

Voluntary participation in major international institutions and compliance to established rules in the international community are essential for soft power maintenance. International institutions are "the rules that govern elements of world politics and the organizations that help

[11] Zhang Zhihong, "The Forces Behind China's Climate Change Policy", in Paul G. Harris (ed.), *Global Warming and East Asia* (London: Routledge, 2003), p. 78.

implement those rules."[12] Although the international society is still described as anarchical with world politics driven by the traditional exercise of state power, international institutions are increasingly important and have become less marginalized. In reality, even the most powerful states have been relying heavily on international institutions — not so much the UN as other organizations and regimes have set rules and standards to govern specific sets of activities.[13] Institutions create the capability for states to cooperate in mutually beneficial ways by reducing the costs of making and enforcing agreements — what economists refer to as "transaction costs."[14] Therefore, a state's position on various international institutions is extremely important for its soft power expansion, because this directly affects its moral high-ground in the eyes of other states that value cooperation in the framework of these institutions. The soft power of a country rests primarily on three resources: its culture, its political values, and its foreign policies.[15] Whether to join an international institution is not simply a part of a country's foreign policies, but also an indicator of its political values.

Case in point, the United States showed how one's international image was badly impaired by not participating in the international climate institution. In 2001, US President George W. Bush announced that his country would withdraw from the Kyoto Protocol because the deal did not include meaningful participation by the key developing countries. The United States' unilateral departure from the protocol immediately aroused long-lasting global anger and criticism.

Economic Gains

Huge Lucre from CDM Projects[16]

To attract both developed and developing countries to join the Kyoto Protocol, the protocol introduced a flexible mechanism called the Clean

[12]Robert O. Keohane, "International Institutions: Can Interdependence Work?" *Foreign Policy*, Issue 110 (1998), p. 82.
[13]*Ibid.*, p. 84.
[14]*Ibid.*, p. 86.
[15]Joseph S. Nye, *Soft Power: The Means to Success in World Politics* (New York: Public Affairs, 2004), p. 11.
[16]For details concerning China's benefit from the CDM projects, please refer to Chap. 10.

Development Mechanism (CDM). The CDM encourages industrialized nations to give more fiscal and technological aid to developing countries for environmental protection purposes. As the deadline (year 2012) for industrialized nations' 5% emission reduction is approaching, they are in urgent need to buy the CERs from developing countries to meet the Kyoto requirement. Even if some of them are able to fulfill the goal on their own, they still have the incentives to buy CERs because the emission reduction cost in developing countries is much lower than that in developed countries.

China, the largest developing country with one of the world's largest GHG emissions, enjoys the highest CDM project potential because its marginal cost for emission reduction is relatively low compared to many other developing countries. Experts attribute this to its low technological level and the use of inefficient energy equipment. It is often more difficult and expensive for a project with high levels of technology and advanced facilities to reduce emissions, as the room for further GHG abatement is already limited.

On 30, June 2004, 7 months before the implementation of the Kyoto Protocol, the Chinese government enacted a draft regulation on the management of domestic CDM projects, showing that it was prepared and eager to benefit from the protocol. Although China consumed less than two-fifths of the developing world's fossil fuel consumption (the major source of GHGs), China's CDM project potential probably could account for half of the world's total. The number of China's CDM projects registered at the United Nations lags behind that of India, but the actual CERs produced in China annually far exceed those of India — accounting for about half of the world's total. By 13, May 2008, a total of 1295 CDM projects had been approved by the Chinese government.[17] The average selling price for each contracted CER in China thus far has been about US$ 7. One report said in 2006, China secured contracted CER income of US$ 3 billion for dozens of CDM projects, about 60% of the world's total value of carbon credit sold.[18]

---

[17]National Development and Reform Commission (NDRC), *Latest Development of CDM Projects Approved by the NDRC*, Clean Development Mechanism in China (3, May 2008), http://cdm.ccchina.gov.cn/WebSite/CDM/UpFile/File1809.pdf.

[18]Keith Bradsher, "China Knows Which Way Wind Blows", *International Herald Tribune* (10, May 2007), p. 1.

Optimizing Economic Structure and Accelerating Technology Upgrade

The rapid economic growth, coupled with the accelerated industrialization and urbanization process in China, has imposed huge resource and environmental pressure upon the country's feeble ecological systems. With a population of 1.3 billion, China tried to become a world manufacturing center after its accession to the World Trade Organization (WTO) in 2001. When domestic consumption was determined to be insufficient to support economic growth, the country relied heavily on its rocketing export policy. Many products exported from China are low value-added or energy-intensive, which consume huge resources and exacerbate the environmental problems in the country. Since 2003, China's GDP growth accelerated to more than 10% annually, with its output in iron and steel, cement, aluminum, and other heavy and chemical industry soaring, and fixed asset investment becoming overheated. In 2002, China's total energy consumption was comparable to that of 1997, with the annual energy consumption growth much lower than the GDP growth during the same period. Since then, however, growth in energy consumption has been outpacing economic growth by large margins. After Wen Jiabao's ascension to premiership in 2003, he began to worry about this unreasonable economic restructure that put huge pressure upon the country's natural environment and caused bottleneck supply in power and raw materials. Despite a series of measures (the so-called "macro-control" policies) since 2004, the overheating problem could not be effectively alleviated; concurrently, the shortage of natural resources compounded by an increase in environmental pollution exacerbated an already alarming trend.

The implementation of the Kyoto Protocol in 2005 provided a rare opportunity for the country to restructure its economy from an energy-intensive heavy industry into a low-carbon model. The protocol includes commitment from industrialized nations to increase technology transfer and fund assistance to developing countries for emission-cutting activities. Economic structure is closely related to a country's science and technological development level, and the import of various green technologies, such as clean coal, energy-efficient building, and renewable energy development, will help the country to accelerate the economic restructure and industrial upgrade. Over 1000 CDM projects in the country will experience a ripple effect, spreading environmental protection technologies and green production ideas on a grander scale. Due to the infusion of foreign investment and advanced technology in the implementation of CDM projects, scholars expected CDM programs to boost the country's GDP by 0.03% in 2010, 0.34% in 2020, and 0.52% in 2030.[19]

Easing Energy Security Anxiety

The enactment of the Kyoto Protocol and future international climate mechanisms will assist China to alleviate its anxiety about energy security issues to some extent. First, energy-efficiency programs and support of renewable energy will curb China's rocketing demand for fossil fuels. China has abundant coal reserves but is poor in oil and natural gas reserves. Increased reliance upon imported petroleum and natural gas has made China determined to tap renewable energy such as hydro, solar, wind, and thermal power for alternatives. In its 11th Five-Year Plan, the government has vowed to increase the ratio of renewable energy in total energy consumption to 10% by 2010.

Second, China can also benefit from the protocol's role in repressing international petroleum and natural gas prices when the world is trying to cut carbon emissions. Industrialized countries' emission reduction obligation under the KP framework is certain to affect international energy prices because OECD countries have to restrain their domestic petroleum, natural gas consumption, and reduce energy imports. The drop in international petroleum and natural gas prices caused by shrinking fossil fuel demand after the implementation of the KP will help China save money in importing oil and gas. With increasing demand for imported petroleum and natural gas, China is very sensitive to international energy price fluctuations. During the 6 years from 2001 to 2006, China's petroleum import volume soared 432%, making it the world's third largest petroleum importer after the United States and Japan. The emission reduction obligation committed by many OECD countries will help alleviate China's anxiety about

[19] Lu Xuedu and Liu Deshun (eds.), *Clean Development Mechanism in China: Taking a Proactive and Sustainable Approach* (Beijing: Tsinghua University Publishing House, 2004), p. 133.

international energy price hike. The World Bank has estimated that Kyoto Protocol's emissions reduction target would bring down petroleum prices by 10% and natural gas prices by 17% in 2010, if emission trading is not allowed and other factors are not considered.[20] Even if emission trading is carried out among industrialized countries, a reduction of petroleum and natural gas prices by 5% and 14%, respectively can still be expected.[21]

Improving Domestic Governance

Environmental pollution is a global issue that most fast-growing economies have to face: relating to many factors such as economic structure, technological level, political systems, governance capacity, institutional building, as well as public awareness and social participation. In China's state-centric society and one-party political system, the government's capacity building ability becomes extremely important for handling those "growing pains."

Since the early 1990s, China's officialdom started to measure local officials' performances and granted promotions based on GDP growth figures. Some attribute ecological degradation to this GDP-first assessment system that often ignores environmental protection. Pressured by its commitment to the Kyoto Protocol to mitigate climate change, provide national emissions inventories, report on national green programs, and promote information exchange, the Chinese government has realized that it is time to change the GDP-only mindset and embrace green indices when evaluating the officials' performances.

In November 2007, China's State Council released a circular that uses green figures of per-unit GDP energy consumption and major pollutants reduction to assess provincial officials' performances.[22] In lieu of pure economic figures, the Chinese government is trying to measure local officials'

---

[20]Ellerman, A. Denny, Jacoby, Henry D. and Decaux, Annelene, "The Effects on Developing Countries of the Kyoto Protocol and CO_2 Emission Trading", The World Bank Policy Research Working Paper Series (31, December 1998), p. 15, http://www-wds.worldbank.org/servlet/WDSContentServer/WDSP/IB/2000/02/24/000094946_99031911110541/Rendered/PDF/multi_page.pdf.

[21]*Ibid.*

[22]State Council, *Implementation Plan for Unit GDP Energy Consumption Assessment System*, The Central People's Government of the People's Republic of China (17, November 2007), http://www.gov.cn/zwgk/2007-11/23/content_813617.htm.

achievement through a much more complicated system that also includes environmental indices. According to the new regulation, major officials of the provincial-level regions as well as leaders of the top-1000 state-owned or controlled energy consumption enterprises are to be assessed by per-unit GDP energy consumption figures calculated through quantified scoring methodology.

Meanwhile, after the Kyoto Protocol took effect in 2005, China had more frequent exchanges on climate change programs with international organizations such as the World Bank and the United Nations Development Program (UNDP), as well as the government of Italy, Canada, France, Japan, and Germany, which helped central and local Chinese government to establish emission reduction service centers for capacity building. The Chinese government is now conducting more training activities for officials, senior managers, and consultants to familiarize them with the procedures of environmentally-friendly projects and to utilize this opportunity to absorb foreign fund and technology.

"DEVELOPMENT FIRST" AND CHINA'S INTER-AGENCY MECHANISM ON CLIMATE CHANGE

Despite Chinese government's increasing attention to environmental issues and the urgency of climate change, the country's "development first" mindset will basically remain unchanged and continue to shape its position in the international climate negotiations. The ruling party still needs economic growth to legitimize its governance, and the current inter-agency bureaucracy on climate change guarantees such direction.

National Climate Change Program's Focus on Development

China publicized a lengthy national program in response to global warming on 4, June 2007. The "National Climate Change Program", which took the Chinese government 2 years to formulate, documents the efforts China has made in tackling climate change, analyzes the future challenges for China, and describes China's policies and positions on international cooperation.

Climate change "is an issue involving both environment and development, but it is ultimately an issue of development,"[23] the national program, Beijing's first comprehensive policy document on the issue, stated in its first paragraph. This indicates that China will continue to put economic growth ahead of climate change without committing itself to quantified emissions cutting.

However, this does not mean that China's original stance on climate change remains unchanged. In contrast to its previous skepticism about the certainty of global warming trend, the national program uses much stronger language to confirm the trend and describe the serious consequences of climate change. China admitted in its national program that "annual average air temperature has increased by 0.5°C to 0.8°C during the past 100 years, which was slightly larger than the average global temperature rise."[24]

By publicizing the national program, China wanted the international community to be aware that the Chinese government pays intensive attention to the issue, but unique national conditions, such as coal-dominated energy structure, prevent the country from accepting mandatory emissions reduction in the current stage.

Despite refusal to shoulder quantifiable emissions reduction, China's national program was putting forward some voluntary targets in energy conservation and fuel switching that can be converted to certain emissions cutback. The plan reiterated the goal in China's 11th Five-Year Program that the country will achieve the target of 20% reduction of energy consumption per unit GDP by 2010 (Table 9.1). This single national target could lessen China's emissions by about 1.5 billion tons by 2010.[25]

The release of the national program came at a calculated time so as to fend off international pressure. On the same day the program was made public, the Ministry of Foreign Affairs announced that President Hu Jintao was expected to discuss China's views and policies on global climate change during his meeting with the G-8 leaders in Germany.

---

[23]National Development and Reform Commission (NDRC), *China's National Climate Change Program*, China Climate Change Info-Net, June 2007, p. 2, http://www.ccchina.gov.cn/WebSite/CCChina/UpFile/File188.pdf.

[24]*Ibid.*, p. 4.

[25]Joanna Lewis, "China's Strategic Priorities in International Climate Change Negotiations", *The Washington Quarterly*, Vol. 31, No. 1 (2007), p. 167.

Table 9.1. Estimated Emission Reductions for Select Energy Saving Measures by 2010.

Emission Reduction Means	Estimated Emission Cutting (million tons of CO_2 equivalent)
Developing hydropower stations especially in western areas	500
Developing nuclear power stations	50
Upgrading technology in thermal power generation	110
Developing coal-bed methane, coal-mine methane industry	200
Developing bio-energy	30
Developing wind, solar, geothermal, and tidal energy	60

Source: National Development and Reform Commission (NDRC), *China's National Climate Change Program*, Clean Development Mechanism in China (28, February 2007), pp. 32–34, http://cdm.ccchina.gov.cn/english/NewsInfo.asp?NewsId=1559.

Inter-Agency Mechanism on Climate Change

China's "development first" stance is closely related to its special inter-agency mechanism on climate change in which the National Development and Reform Commission (NDRC) plays a dominant role, while the State Environmental Protection Administration (SEPA), the predecessor of the current Ministry of Environmental Protection, and China Meteorological Administration (CMA) play minor roles.

Starting in the 1980s, China treated climate change as a scientific issue and gave the CMA the responsibility of advising the government on policy options in the international arena involving the UNFCCC.[26] As political awareness and sensitivity surrounding climate change increased in the late 1990s, this role shifted to the more powerful State Development and Planning Commission, which has since evolved into the NDRC.[27] In 1998, China established the National Coordination Committee on Climate Change, an inter-agency organization that later included representatives from the NDRC, Foreign Ministry, Ministry of Science and Technology, CMA, SEPA, and other 10 ministerial-level departments.

[26]Joanna Lewis, "China's Strategic Priorities in International Climate Change Negotiations", *The Washington Quarterly*, Vol. 31, No.1 (2007), p. 158.
[27]*Ibid.*

With the NDRC director appointed by the Chairman of the National Coordination Committee on Climate Change, the NDRC has been managing all the routine affairs of the Committee and clearly has the pivotal role in the decision-making process. Compared with SEPA[28] and CMA (potentially sympathetic to the more pro-active approach of emission abatement), the NDRC leans to rapid economic growth as well as materialized benefits including fund and technology transfers. The main function of the NDRC, a macro-economic regulatory department, is to maximize economic development. From its inception, the Commission's position was very skeptical toward scientific concerns regarding the causes and impacts of climate change, and it opposed any commitments not in line with economic growth, e.g. GHG emission reductions.[29]

Inter-agency conflicts between the NDRC and SEPA have become problematic as domestic environmental issues have increased. The latest evidence of such inconsistent behavior is China's car fuel policy gone awry. Vigorously pushed by SEPA, the central government had planned, on 1, July 2007, to introduce a standard for cars known as Euro III, which would reduce carbon monoxide and other emissions. The NDRC, however, in 2007, proposed that the rules should be delayed for up to 2 years because of the lack of compatible fuel — given that the changes could damage new engines.

The high-powered role of the Foreign Ministry and Ministry of Science and Technology enhances NDRC's advantageous position in the climate change decision-making process. The Foreign Ministry puts an emphasis on the preservation of national sovereignty, meaning that developing countries have the need and the right to develop, while the Ministry of Science and Technology focuses on foreign technological aid. Without any change to the inter-agency climate change mechanism, the "development-first" views of the NDRC and Foreign Ministry will undoubtedly continue to shape the major features of China's negotiating position in international talks.

---

[28]SEPA was the fifth important agency in the pecking order within the National Coordination Committee on Climate Change, behind the NDRC, Foreign Ministry, Ministry of Science and Technology and CMA.

[29]Carmen Richerzhagen and Imme Scholz, "China's Capacities for Mitigating Climate Change", *World Development*, Vol. 36, No. 2 (2007), p. 318.

Domestic Pressure and China's Initiative

China admitted that it is facing a potentially devastating impact from climate change, which has become a national, social, and economic problem. The prolonged droughts will lead to a decline in agricultural output, while the acceleration of glacier melting on the northwestern Qinghai–Tibet plateau will reduce water supply to the Yangtze and Yellow Rivers, China's two longest rivers. Furthermore, China's rising sea level triggered by global warming is higher than the global average, and this will threaten China's coastal cities — the nation's economic powerhouses.

The Chinese government has now come to realize that if the worsening environmental problem is not addressed soon, the consequence may be so disastrous as to cancel the benefits derived from the economic boom. The government has been especially sensitive with respect to the heightened public complaints about the pollution. The number of mass disturbances caused by environmental pollution in China has been growing at an annual rate of 30%, according to the former Chinese Vice Premier Zeng Peiyan at a conference on environmental protection. Examples from Eastern Europe and the former Soviet Union show that growing environmental discontent often served as a catalyst for broader opposition to their respective communist regimes.

Wen Jiabao, China's premier, delivered a scathing assessment in April 2007 stating that the government's inability to meet long-stated goals to cut pollution and energy use must be corrected to sustain growth. He said the government will continue to curb the energy-guzzlers by further adjusting export rebates, levying more export tariffs and reducing export quotas.

Anxiety about the gloomy climate prospect has made the public doubt the country's industrial strategy that puts priority on the auto manufacturing sector. In 2006, China became the world's second largest car market and the third largest car producer after the United States and Japan. More than 1000 new private cars are hitting the roads everyday in Beijing, where transport already accounts for 20% of the society's total energy consumption. Frequent traffic congestion exacerbates the problem, making these vehicles less efficient with more emissions when queuing on the streets. More than 100 Chinese cities including Beijing and Shanghai took part in the country's first official urban "car-free day" on 22, September

2007, barring automobiles from selected areas and ordering officials to swap their black sedans for public transport.

It is the most opportune time for the central government to restructure its economy to wean away from heavy industry including steel, cement, and aluminum. The battle has begun between Beijing and provincial/city authorities over implementing the central government's energy and pollution standards. Due to both internal and external pressures, the central government has initiated a series of new policies and actions to better address the climate change and energy conservation issues.

In 2007, China established a high-level group on climate change and energy-saving headed by Premier Wen Jiabao himself, with then Vice Premier Zeng Peiyan and State Councilor Tang Jiaxuan as his assistants. China's Foreign Ministry later announced that it had also established a leading group in charge of international climate change affairs, headed by Foreign Minister Yang Jiechi. The Chinese government wanted to use these high-profile initiatives to show the world its serious attention to the climate issue and mobilize its cadres who for a long time ignored the problem. The newly-established National Leading Group on Climate Change, Energy Saving and Emission Cutting is expected to emphasize more on the environmental aspect of the cabinet's policy-making process, but the NDRC's strong position will not be modified much as the National Leading Group is housed within the NDRC.

In November 2007, China launched a state-owned CDM fund to finance the country's efforts to address climate change. The China CDM Fund, managed by the Ministry of Finance, obtains money from the current CDM projects that help China improve energy efficiency and protection of the environment. The Fund also receives donations from financial institutions such as the World Bank and the Asian Development Bank. The purpose of the Fund is to promote more projects involving clean energy, energy conservation, and emission reduction in China.

China wants to improve its international image on environmental protection through bilateral or multilateral green projects, with the landmark China–Singapore "Eco-City" Project being the most high-profile venture. According to the agreement signed in November 2007, the project located in the Binhai New Area of Tianjin Municipality would use advanced energy-saving and environmental technologies in the hopes of turning Tianjin into a model "green" city. In cooperation with Britain, China is building another low-carbon "Eco-City" in Dongtan, a town near Shanghai City. The government hopes that these projects' management, technology, and policy experience will serve as a model for other Chinese cities.

CHINA'S FUTURE STANCE IN POST-KYOTO NEGOTIATIONS

The Fourth Assessment Report (AR4), released by the Intergovernmental Panel on Climate Change in 2007, describes the connection between climate change and human activity with a high degree of certainty. The Kyoto Protocol, which is scheduled to expire by 2012, cannot meet the demand from the international community for further emission reductions. At the UN climate change summit in Bali (Indonesia), December 2007, a new round of international negotiations was formally introduced to work out a new accord substituting the Kyoto Protocol. In the post-Bali talks, environmental activists are hoping that the United States plus China and India will all be included in the new treaty with quantitative emission quotas or even a comprehensive carbon tax. China is in a dilemma. On the one hand, it is fretting about reputational damage precipitated by its deteriorating environmental problem and rapidly-growing emissions, banking on international cooperation to solve it; on the other hand, it fears that emission abatement obligations may block its fast economic growth. In future talks, it is still very unlikely for China to yield to international pressure and accept mandatory emission-cutting targets, but it will probably use voluntary carbon-intensive benchmarks for substitution of cap-and-tax obligations. Moreover, China will participate more frequently in other bilateral and multilateral diplomatic activities on climate change, aside from the UN-track climate talks, to polish its international image and alleviate external pressures.

Voluntary Targets Instead of Absolute Emission Obligations

Regarding compulsory emission targets as a cap on domestic economic growth, Chinese officials have on many occasions articulated the country's adherence to the "common but differentiated responsibilities" and objection to any legal responsibility on quantifiable reductions. In their press conferences during the annual session of the National People's Congress (NPC) in 2007, both Chinese Premier Wen Jiabao and Foreign Minister Li Zhaoxing were asked by the European media about China's responsibility on the climate change issue. Neither of them made any commitment to shoulder legally-binding obligations for emission reductions. At a news conference on 22, November 2007, less than one month before the Bali climate summit, Chinese Foreign Ministry spokesman Liu Jianchao stated developing countries should not have to meet compulsory targets set by rich nations.[30] During the Bali conference, Su Wei, a top climate expert for the government and member of the Chinese delegation, stated it was unfair to ask developing nations to accept binding emissions cuts and other restrictions being pushed by industrialized states.[31]

From China's perspective, international effort against climate change is a double-edged sword: while a positive outcome in terms of fund and technology transfer to China, a wholly different and negative impact if the shackles are placed upon the giant from a disadvantageous pact arising from future negotiations. Chinese experts argue that if China accepts emission quotas similar to that of the industrialized nations in the Kyoto Protocol, the cost will outweigh any current gains derived from CDM projects to buy emission permits from other countries at a higher carbon price in the post-Kyoto period. The alternative carbon tax, if adopted in the next protocol, is expected to cost the Chinese industry billions of dollars or more — given China's economy and emissions growth to continue at the current rates.

China's negotiation tactic is clear: utilize voluntary carbon-intensity-based targets based on the reduction of energy consumption or emissions per unit of economic output to replace mandatory emission cuts and ease international pressures. As a result, such a target is inherently indexed to

[30]"China: Developing Nations Should Set Own Emissions Reduction Targets", *Xinhua News* (22, November 2007).
[31]"China Rejects Climate Mandates, Blames Lavish US for Crisis", *AP News* (7, December 2007).

the economic growth of a country, and meeting this target does not directly require a decrease in economic production; therefore, this type of target is more palatable to developing countries.[32] The main limitation of an intensity-based target is that, although it can lower an emission growth trajectory below the projected "business-as-usual" level, it is unlikely to result in an absolute decrease in emissions.[33] As climate change engender imminent foreseeable disastrous consequences, such moderate goals cannot satisfy many environmentalists and negotiating parties. The European Union passed a new Energy Plan in March 2007, requesting China, India, and the United States to sanction a more ambitious plan to cut GHG emissions.

Acknowledging the international pressure for its compulsory emissions reduction, the Chinese government is trying to play a low-profile role in the global carbon trade. In February 2007, several reports surfaced intimating that Beijing hoped to establish a bourse for trading carbon emissions permits to capitalize on China's emergence as a hotspot in the multi-billion dollar carbon market. China's Office of the National Coordination Committee on Climate Change posted a statement on its web-site that stated the contrary. The short statement stated — "currently the Chinese government has no plan to establish a so-called 'Climate Exchange' " — adding that under the Kyoto Protocol, such emissions trading schemes could only be set up in countries with binding targets.[34]

Continued Solidarity with Developing Countries

More than 100 developing countries have been acting collectively since the early days of the global climate negotiations to strengthen their voices and fend off pressures from industrialized nations. Although huge political, economic, and geographic disparity have always existed among the

[32]Joanna Lewis, "China's Strategic Priorities in International Climate Change Negotiations", *The Washington Quarterly*, Vol. 31, No.1 (2007), p. 166.
[33]*Ibid.*
[34]China's Office of the National Coordination Committee on Climate Change, "Comment on the Report about Setting Up a Climate Exchange in China" (2007), http://cdm.ccchina.gov.cn/web/main.asp?ColumnId=25.

Third World, they nonetheless formed the "G-77 plus China" bloc to build a common purpose and safeguard collective interest when negotiating with the affluent western countries. Despite its large size and special status in the UN Security Council, China has preferred to ally itself with G-77 to deal with the complicated and new topic of climate change that was first proposed by the industrialized world. In the post-Kyoto negotiation that was officially launched during the Bali conference, China continued to enhance strategic solidarity with the Third World rather than deviate from the bloc and act alone. In fact, China's alliance with the G-77 has not waned, and its willingness to "go-it-alone" has declined even further. With increasing economic growth and energy use, fear of being singled out grows ever more likely.[35]

China's desire to stay inside the bloc, however, is facing more challenges due to some members' willingness to break ranks, while others are complaining about the unequal distribution of CDM profits among developing countries. Some Latin American countries, led by Brazil and Argentina, have shown eagerness to accept similar quantitative emission obligations as western countries, while China and India fear legally-binding goals will hamper their burgeoning industrialization process. Small island countries that are more vulnerable to climate change are expressing their dissatisfaction about responsibility shirking by major emitters, while petroleum-exporting nations such as Saudi Arabia pose as strong objectors to the international climate regime.

Meanwhile, the imbalance of CDM income distribution as well as China's surging energy consumption and emissions have led to many developing countries' discontent about the current climate change regime dominated by the Kyoto Protocol. Many developing countries, including mostly poorer nations vulnerable to global warming, have been largely left out of the wealth-sharing of CDM projects. In 2006, all of Africa received less than US$ 150 million from the sale of carbon credits from CDM projects, or 3% of the global total. With the foreign exchange reserve of more than US$ 1.7 trillion, China is no longer a poor member of the Third World.

[35]Joanna Lewis, "China's Strategic Priorities in International Climate Change Negotiations", *The Washington Quarterly*, Vol. 31, No.1 (2007), p. 166.

Whether other developing countries will still regard China as one of their own and continue the close alliance is questionable.

More Attention to Regional Mechanism

As the UN-track (or Kyoto-track) climate change regime is featured with quantifiable emission-cutting targets, China is planning to join more regional mechanisms on climate change that emphasize noncompulsory energy-efficiency programs. Explicitly, China still claims that global climate talks should be conducted in the framework of the United Nations, but in practice, China has become more active than before in joining all types of bilateral and multilateral cooperation mechanism against climate change — outside the purview of the UN-track.

Multilaterally, China is more willing to accept international agreements that call for voluntary actions rather than those with mandatory emission-cutting targets. This will undoubtedly assist China to enhance its soft power while protecting its hard power (economic growth). In 2005, China joined the collective action with Australia, India, Japan, South Korea, and the United States to establish the Asia-Pacific Partnership on Clean Development and Climate, which emphasized energy conservation and technology cooperation in lieu of setting mandatory emission-cutting targets. The United States and Australia, two of its members, refused to ratify the Kyoto Protocol at that time. The Asia-Pacific Partnership on Clean Development and Climate is a cooperation mechanism outside the UNFCCC and Kyoto Protocol; as a member, China obtains advanced energy-efficiency technology from affluent countries, coordinates diplomatic positions with other heavyweights in international climate talks, and improves its international image in the green area without accepting legally-binding targets. In recent years, many existing multilateral organizations, including Asia-Pacific Economic Cooperation (APEC), Association of Southeast Asian Nations (ASEAN) Plus Three (China, Republic of Korea, and Japan), and G-8, have started to augment special mechanisms on climate change. Subsequently, China has proved to be an active participant or dialogist in forums, conferences, and cooperation programs held by these mechanisms. During the September 2007 APEC summit in

Sydney, both China and the United States accepted a program with non-binding goals on energy efficiency and afforestation.[36]

Bilaterally, China is attentive to cooperative programs on clean energy and emission reduction with industrialized nations including the United States. Climate change issue has become an indispensable area of interest within China's recent bilateral talks with many western countries. China also enhances bilateral coordination and cooperation with India and other developing powers on the climate issue. During the Indian Prime Minister Manmohan Singh's visit to China in January 2008, climate change was also on the agenda and the two sides reached consensus that neither China nor India should be obligated to make commitments to reduce carbon emissions.

China will definitely face greater challenges and pressures in the coming climate talks. China's economic strength (hard power) has grown so rapidly in the past decade that the international community expect it to shoulder greater international responsibilities on climate change. While unduly problematic for China to accept legally-binding targets for emission reductions, state leaders understand very clearly that its soft power will be greatly damaged if it continues to pursue conventional means. China will try to gain more support and sympathy from both developed and developing countries through frequent diplomatic exchanges and work to reduce domestic carbon emissions per GDP unit continuously and remarkably — with the hope to fulfill zero carbon growth by the middle of this century. China would need to rethink the current tactics if the United States were to reverse its position and take on an emission reduction commitment. Predictably, even under this scenario, China's development-first mindset and inter-agency bureaucracy would still make legally-binding obligations unacceptable for China.

---

[36] Asia-Pacific Economic Cooperation (APEC), "Sydney APEC Leaders' Declaration on Climate Change, Energy Security and Clean Development", Asia-Pacific Economic Cooperation 2007 Leaders' Declaration (9, September 2007), http://www.apec.org/apec/leaders_declarations/2007/aelm_climatechange.html.

According to the program, APEC members will reduce "energy intensity" — the amount of energy needed to produce a dollar of gross domestic product — by 25% by 2030. They pledged to increase forest cover in the region by at least 20 million hectares (50 million acres) by 2020.

Chapter 10

How China Manages and Benefits from CDM Projects

The Kyoto Protocol, the first international treaty to set detailed obligations for industrialized nations to reduce their greenhouse gas (GHG) emissions by 2012, is in reality yielding huge benefit for China instead of imposing afflictive restraint upon the world's fastest-growing economy.

China, then the world's second largest GHG emitter after the United States, completed its domestic ratification process in 2002, and appealed to other hesitating parties to ratify the protocol as soon as possible. The Kyoto Protocol came into effect in 2005, requiring 38 industrialized countries to reduce their GHG emissions between 2008 and 2012 by an average of 5.2% below 1990 levels.

China is NOT one of the 38 countries named, and it is not required to reduce GHG emissions. With no compulsory burden to shoulder, China actually is economically incentivized by the Kyoto Protocol's Clean Development Mechanism (CDM) to participate in the international joint actions against climate change, announcing again and again that the future international climate talks should continue to follow the Kyoto tracks.

CHINA'S HUGE POTENTIAL IN CDM PROJECTS

To attract both developed and developing countries to join the Kyoto Protocol, the protocol introduced a flexible mechanism called Clean Development Mechanism (CDM). For industrialized nations such as Japan and Canada which are having difficulties meeting even the minimum requirement set by the protocol, they can invest in overseas environmental projects in developing countries that have no emission reduction obligations and then buy the certified emission reductions (CERs) to offset their obligations.

CDM and Foreign Investment Inflow

The CDM encourages industrialized nations to give more fiscal and technological aid to developing countries for environmental protection purposes. As the deadline (year 2012) for industrialized nations' 5% emission reduction approaches, they are in urgent need to buy the CERs from the developing countries to meet the KP requirement. Even if some of the nations are able to fulfill their emission reduction goal through projects within their own country, they still have the incentives to buy CERs because the emission reduction cost in developing countries is much lower than that in developed countries.

China, the largest developing country with the world's second largest GHG emissions, enjoys the highest CDM project potential because its marginal cost for emission reduction is relatively low compared to many other developing countries. Experts attribute this to its low technological level and the use of inefficient energy equipment. It is often more difficult and expensive for a project with high standards of technology and facility to cut down emissions as the room for further GHG abatement is already limited. Take India as an example; its emission reduction cost per ton carbon equivalent is about US$ 34.1, while that in China is only about one-third or US$ 12.67.[1]

The World Bank estimated that China's CDM project potential could account for half of the world's total. The Organization for Economic

[1] Cui Dapeng, "Analysis of International Negotiation on Climate Change from the Perspective of Political Economy", Doctoral dissertation, Beijing: Renmin University of China (June 2002), p. 100.

Table 10.1. Net Growth of Foreign Investment Caused by CDM in China (US$ 1 Million).

Year	Net growth of foreign investment
2005	238
2010	475
2015	356
2020	238

Source: Lu Xuedu and Liu Deshun (eds.), *Clean Development Mechanism in China: Taking a Proactive and Sustainable Approach* (Mandarin Edition), (Beijing: Tsinghua University Publishing House, 2004), p. 133.

Co-operation and Development (OECD) also said in its report that most CDM projects would best be operated in China.[2]

Besides fund injection and technology transfer in CDM projects, both the Chinese corporations and government can gain huge profit from the sale of CER to foreign organizations. The UN predicts that through CDM projects, China will become the largest carbon credits provider in global carbon trade by 2012, and experts expected CDM programs to increase foreign investment in China by US$ 475 million in 2010 (about 0.7% of current annual foreign direct investment in China), US$ 356 million in 2015, and US$ 238 million in 2020 (Table 10.1).[3]

Although this increase in foreign direct investment is only a small portion compared to the tens of billions of dollars China attracts each year, the CDM development is a surprising windfall that may have the catalytic effect of replacing outdated installations, promoting public awareness about environmental protection and solving the pollution problems efficiently. In fact, the real figure last year was beyond expectation. According to the figures from the World Bank, China captured US$ 3 billion of the world's US$ 4.8 billion in subsidies in 2006 for dozens of CDM projects.[4]

[2]Organization for Economic Co-operation and Development (OECD), Action Against Climate Change: The Kyoto Protocol and Beyond (France: OECD, 1999), p. 56.

[3]According to China's Ministry of Commerce, total foreign investment in 2006 was US$ 69.5 billion.

[4]Keith Bradsher, "China Knows Which Way Wind Blows — Country Gets Largest Share of UN Credits to Curb Emissions", *International Herald Tribune* (10, May 2007).

Chinese Government's Revenue from CDM Projects

On 30, June 2004, 7 months before the implementation of the Kyoto Protocol, the Chinese government enacted a draft regulation on the management of CDM projects in China, showing that it was prepared and eager to gain from the Kyoto Protocol. On 12, October 2005, with the amendment of the draft regulation, the National Development and Reform Commission (NDRC), the Ministry of Science and Technology, the Ministry of Foreign Affairs, and the Ministry of Finance jointly issued the Measures for Operation and Management of Clean Development Mechanism Projects in China to replace the draft regulation.[5]

According to China's Measures for Operation and Management of Clean Development Mechanism Projects in China, revenue from the transfer of CERs shall be owned jointly by the Government of China and the CDM project owner, with the allocation ratio defined as follows:

(a) the Government of China takes 65% CER transfer benefit from hydrofluorocarbons (HFC) and perfluorocarbons (PFC) projects;
(b) the Government of China takes 30% CER transfer benefit from N_2O project;
(c) the Government of China takes 2% CER transfer benefit from CDM projects in priority areas and forestation projects.[6]

Two CDM Cases

In August 2005, just half a year after the official implementation of the Kyoto Protocol, Japan's JMD Greenhouse Gases Reduction Co., Ltd. and the Juhua Co., Ltd. based in East China's Zhejiang Province agreed to establish a CDM project to help the Chinese firm reduce nearly 40 million tons of its carbon dioxide equivalent discharges over 7 years. The JMD provided all funding and technologies to decompose HFC23 discharged

[5]Office of the National Coordination Committee on Climate Change and Institute of Nuclear and New Energy Technology of Tsinghua University, *Guidelines on China's CDM Project Development* (Beijing: China Environmental Science Publishing Company, 2006), p. 55.

[6]China's Measures for Operation and Management of Clean Development Mechanism Projects in China, Article 24, China Climate Change Info-Net (18, October 2005), http://www.ccchina.gov.cn/cn/NewsInfo.asp?NewsId=3780.

by a chemical plant owned by Juhua. The decomposition of the potent greenhouse gas HFC23 is the so-called "low-hanging fruit" in carbon offset projects. This gas from the perfluorocarbon family of gases is a by-product of the production of refrigerant gases. It is relatively cheap to deal with, simply being burnt. Yet, doing so is a very lucrative activity in global carbon markets because each tonne of HFC23 is equivalent to 11,700 tonnes of CO_2 in its global warming impact. So a small amount of HFC23 destruction earns a huge volume of CERs. According to the contract, the JMD would purchase emission credits at a price of at least US$ 6.5 per ton CO_2 equivalent, which is equal to at least US$ 36.5 million every year or a total of US$ 255.8 million for the seven contractual years based on the anticipated annual reduction of more than 5.62 million tons of CO_2.[7] This became the first large-scale CDM deal in China.

A few months later, another Chinese chemical plant signed an even larger emission reductions deal with the World Bank. By then, this was the largest-ever CDM project in the world. According to the contract, the Changshu 3F Zhonghao New Chemicals Material Co. Ltd. in Jiangsu Province would receive 438 million euros for cutting HFC23 emissions by the equivalent of 10.43 million tons of CO_2 annually for the next 7 years.

These were China's first two large-scale CDM businesses, in which the Chinese government is expected to gain US$ 385 million from selling CERs to international buyers, who can resell the emission permits in the international carbon market to companies and governments that find it difficult to meet the Kyoto Protocol requirement in the 2008–2012 period.

CHINA'S SHARE IN GLOBAL CARBON MARKET

As the first commitment period of the Kyoto Protocol (2008–2012) arrives, the international price for emission permits will probably increase steadily. This possible price increase of carbon credits will probably make China's carbon credit business more lucrative. The supply of carbon credits will be 300 million metric tons short of demand when the Kyoto Protocol on reducing pollution expires in 2012, as forecasted by the World Bank in May

[7]"CDM deal worth 2 billion yuan hitting Juhua", *Shanghai Securities News* (6, January 2006), p. A7.

2007.[8] Increased demand had raised the prices of the credits by more than 70% from February to May 2007, and Japan paid as much as US$ 16 per ton of CO_2 equivalent compared with an average price of US$ 10 in 2006.[9]

The Largest Piece of Cake

World Bank figures show that a total of 450 million carbon credits were sold in 2006 at an average of US$ 10.70 each, representing emissions reductions equivalent to 450 million metric tons of CO_2 (Figure 10.1). China's share of UN carbon emissions credits sold in 2006 was 61%, much larger than that of India, ranking second with a share of 12%.[10] China, the biggest beneficiary that received three-fifths of the money in 2007 from the CDM projects,[11] is no longer so poor. By the end of year 2007, China's foreign exchange reserves hit a record high of US$ 1.53 trillion, topping all the countries in the world.

By 17, September 2007, 790 CDM projects worldwide were registered with the United Nations. Such projects are expected to reduce emissions by 167.6 million tons of CO_2 equivalent every year (Figures 10.2 and 10.3). By then, 114 China-based CDM projects were registered with the United Nations, accounting for 14.43% of the world's total. India witnessed 278 CDM projects registered with the United Nations by then, accounting for 35.19% of the total and topping all the other developing countries in the world. The total CERs produced annually by China-based CDM projects, however, far exceed those from India-based projects, because individual CDM projects in China are much larger-sized than Indian projects in terms of carbon emission reductions. Therefore, although with fewer CDM projects, China actually enjoys a much larger share of CER revenue than India and thus obtains more dollars from the CDM scheme than any

---

[8]Dinakar Sethuraman, "Carbon-credit Sales may Double This Year", *Bloomberg News* (27, May 2007).
[9]*Ibid.*
[10]Keith Bradsher, "China Knows Which Way Wind Blows — Country Gets Largest Share of UN Credits to Curb Emissions", *International Herald Tribune* (10, May 2007).
[11]*Ibid.*

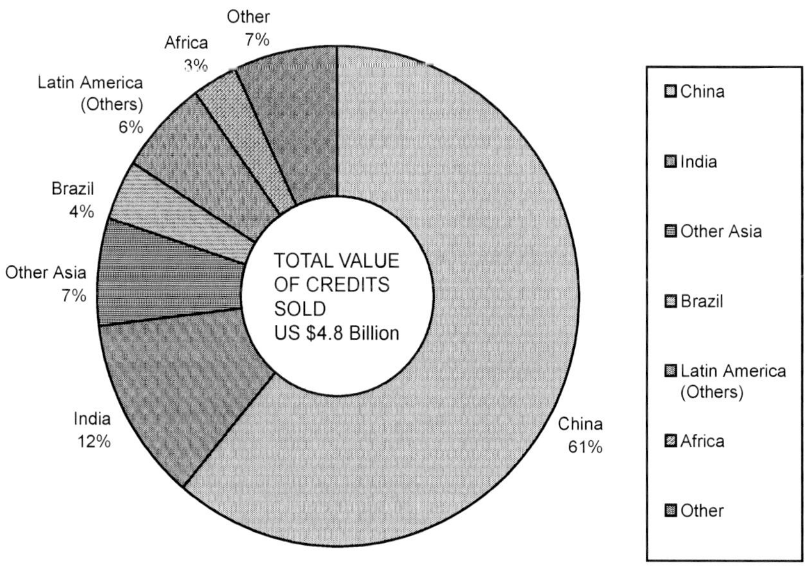

Figure 10.1. Share of UN Carbon Emissions Credits Sold in 2006. (A total of 450 million credits were sold in 2006 at an average of US$ 10.70 each, representing emissions reductions equivalent to 450 million metric tons of carbon dioxide.)
Source: World Bank; Keith Bradsher, "China Knows which Way Wind Blows — Country gets Largest Share of UN Credits to Curb Emissions", International Herald Tribune (10, May 2007).

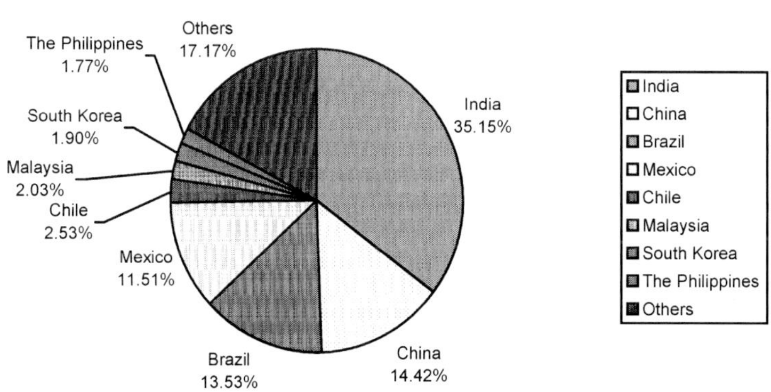

Figure 10.2. Proportion of CDM Projects Registered in UN. (Total Project Number: 790.)
Source: The Statistics are from the CDM statistics from the United Nations Framework Convention on Climate Change (UNFCCC), up to 17, September 2007, http://cdm.unfccc.int/Statistics/index.html.

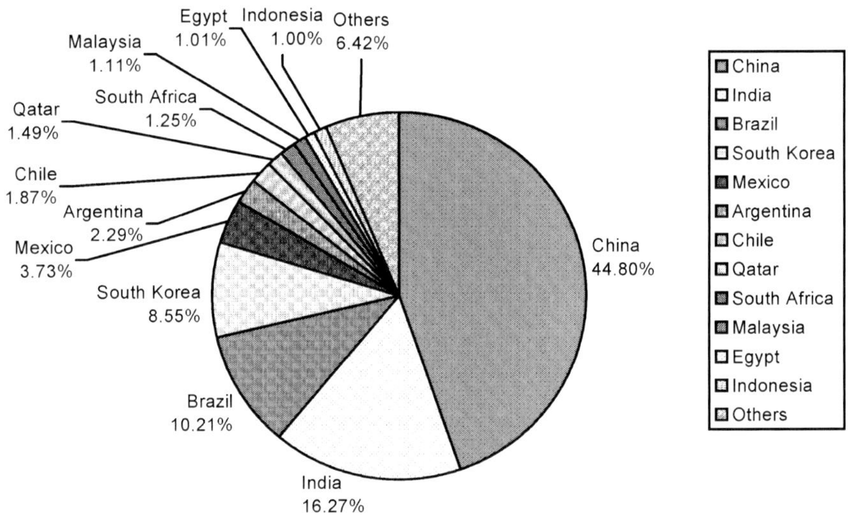

Figure 10.3. Proportion of CERs Produced by UN-Registered CDM Projects Annually. (Total CERs: 167, 602, 438.)
Source: The Statistics are from the CDM statistics from the United Nations Framework Convention on Climate Change (UNFCCC), up to 17, September 2007, http://cdm.unfccc.int/Statistics/index.html.

other country in the world. According to figures provided by the United Nations Framework Convention on Climate Change by 17, September 2007, China's 114 UN-registered CDM projects in reality produce 44.8% of the world's total CERs annually, much more than the 16.27% produced by 278 CDM projects based in India.

Rising Value of Emission Reductions

International carbon trade includes both the project-based emission reduction deals (under the CDM and Joint Implementation schemes stipulated by the Kyoto Protocol) and the direct emission permits transactions (under the Emission Trading scheme stipulated by the Kyoto Protocol). About 90% of the total turnover of the global carbon trading is contributed by the project-based emission deals. Since the Kyoto Protocol took effect in February 2005, the average price of CERs produced by CDM projects has been rising steadily.

From January 2004 to April 2005, the prices of CERs in the international market varied between US$ 3 to US$ 7 per ton of CO_2 equivalent.[12] By the end of 2007, however, the contracted CER prices in some CDM projects had reached about US$ 10 per ton of CO_2 equivalent. Under the European Union Emission Trading Scheme (EU ETS), the largest multinational, greenhouse gas emissions trading scheme in the world, the carbon price has been long time above 20 euro per ton CO_2 equivalent. The EU scheme is largely modeled on the mechanisms in the Marrakech Accords of the Kyoto Protocol, helped by the experience gained during the running of the voluntary UK Emissions Trading Scheme in the previous years.

Contracted CER prices differ not only according to the supply–demand relationship in the international carbon market, but also on the basis of the certainty of CDM projects, which are in most cases financially risky and technologically complicated. According to UN regulations, all CDM projects have to follow the principle of "additionality", which means that the projects must demonstrate that the activities would not have occurred under normal technical and economic conditions in the host country without CER revenues.

The rule of "Economic Additionality" requires that the resources from the Annex I investors are additional to the financial obligations of the Annex I Party under the Convention, and to the current official development aid flows. The principle of "Environmental Additionality" requires that the CDM project activity brings about long-term environmental benefits related to the mitigation of climate change that would not have occurred in the absence of the project.

As CDM project developers' profit from selling CERs to international buyers is additional to the original project revenue, the project's IRR (internal rate of return)[13] will be lifted up with this carbon finance. This means that some environmentally-friendly projects previously regarded as financially-infeasible, suddenly become profitable after being authorized as

[12]Office of the National Coordination Committee on Climate Change and Institute of Nuclear and New Energy Technology of Tsinghua University: *Guidelines on China's CDM Project Development* (Beijing: China Environmental Science Publishing Company, 2006), p. 6.

[13]The internal rate of return (IRR) is a capital budgeting metric used by firms to decide whether they should make investments. It is an indicator of the efficiency of an investment (as opposed to NPV, which indicates value or magnitude).

a CDM project due to the additional revenue through the sale of carbon emission reductions produced by the project.

According to the World Bank's calculation, the IRRs of solid waste management projects can be promoted by 5%–7.5% through becoming a CDM project, while those of collective heating supply projects can be lifted up by 2%–4%.[14] The IRRs of biomass projects can go up by 0.5%–3.5% and the IRRs of wind farm and hydropower projects can increase by 1%–2.5% after they become CDM projects.[15] Experts in China believe renewable energy and energy efficiency projects have the greatest room for expansion in China's carbon market, given China's inefficient energy consumption, favorable government policies, and international technical support. Renewable energy projects are currently not as profitable as traditional power projects in China and elsewhere; so, the CDM will generate extra revenue for these projects thus making new green energy more attractive.

Government's Endeavor to Boost CDM Projects

The Chinese government is now consulting international organizations such as the World Bank and the United Nations Development Program (UNDP), as well as the government of Italy, Canada, France, Japan, and Germany to help local Chinese departments to establish CDM service centers, so as to address the bottleneck of technical services. The Chinese government is conducting more training activities for officials, senior managers, and consultants to familiarize them with the procedures of CDM projects and to utilize this opportunity to absorb foreign fund and technology.

With the assistance of the UNDP, China is establishing a national CDM fund to offer aid for local projects. The NDRC, the Ministry of Finance, the Foreign Ministry, and the Ministry of Science and Technology will jointly set up an auditing council to take charge of the fund. Active CER buyers in the world include international organizations, government-sponsored fund raising organizations, commercial organizations, and nongovernmental

[14]Office of the National Coordination Committee on Climate Change and Institute of Nuclear and New Energy Technology of Tsinghua University: *Guidelines on China's CDM Project Development* (Beijing: China Environmental Science Publishing Company, 2006), p. 7.
[15]*Ibid.*

organizations (NGOs). The World Bank is now the largest CER buyer, managing a number of carbon funds including the Prototype Carbon Fund and the Umbrella Carbon Facility, and acting as a trustee of governmental carbon funds in some European countries. The Umbrella Carbon Facility became a CER buyer in a large-sized HFC23 project in East China's Jiangsu Province.

The British government encourages its business sectors to actively engage in the CDM market.[16] Spain, Sweden, Norway, Austria, Denmark, and Finland have set up agencies for CDM development, allocating a special budget for overseas CDM cooperation. A Japanese Carbon Fund (JCF) was established in December 2004, with a total fund of US$ 140 million. Canada has also established a climate change fund to support the developing countries for their CDM capacity building activities. In 2005, the Canadian government set aside a huge budget for meeting KP requirement, including budget for overseas CDM promotion.

Besides organizations and governments of OECD countries that have legal obligation on emission reduction, other governmental agencies, investment funds, and corporations are possible investors in China's CDM projects and buyers of CERs if they want to profit from the CER resale in the international market. From January 2004 to April 2005, the average CER price in the international market rose by 21% year on year.

According to Chinese regulations, CDM project developers should be Chinese-funded or China-controlled enterprises in China. China's CDM projects are most likely to occur in its agriculture, forestry, coal, steel and iron, power, chemical, transportation, renewable energy, paper making, machinery manufacture, nonferrous metallurgy, construction material, and heat supply industries.

Due to the absorption of foreign investment and advanced technology in the implementation of CDM projects, scholars expected CDM programs to boost the country's GDP by 0.03% in 2010, 0.34% in 2020, and 0.52% in 2030 (Figure 10.4).[17] Although the figure is relatively small in the

---

[16]A CDM promotion section was set up within the Department of Trade and Industry (DTI) of Britain, and in July 2005, DTI organized a delegation to Beijing, Chongqing, and Guangzhou for CDM cooperation and CER purchase.

[17]Lu Xuedu and Liu Deshun (eds.), *Clean Development Mechanism in China: Taking a Proactive and Sustainable Approach* (Mandarin Edition), (Beijing: Tsinghua University Publishing House, 2004), p. 133.

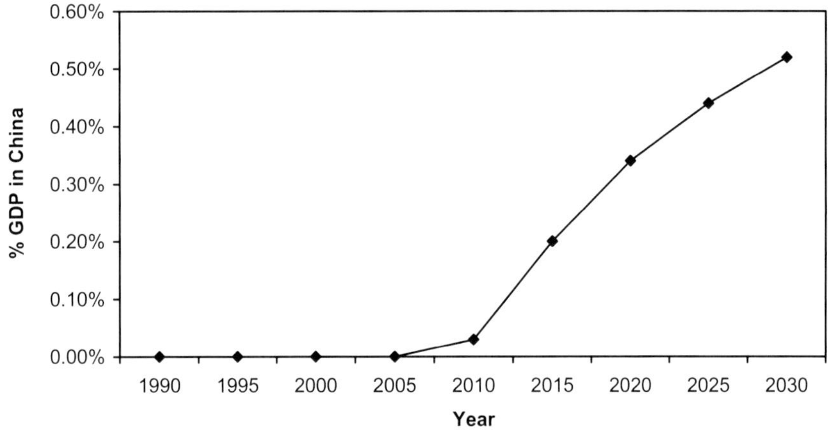

Figure 10.4. Potential Clean Development Mechanism Benefits as % of GDP in China. *Source*: Lu Xuedu and Liu Deshun (eds.), *Clean Development Mechanism in China: Taking a Proactive and Sustainable Approach* (Mandarin Edition), (Beijing: Tsinghua University Publishing House, 2004), p. 133.

preliminary stage, the increase is expected to be accelerated later on because of the ripple effect of environmental projects on the national economy.

In general, the CDM benefits China in the following ways:

- finance emission reduction projects;
- bring China huge profits through sale of CERs;
- provide new channels for environmentally-friendly technology transfer;
- improve energy efficiency and energy conservation;
- improve local environmental conditions;
- alleviate poverty by increasing income and offering job opportunities.

PROCEDURES OF CHINA-BASED CDM PROJECTS

Any CDM project in the world has to be approved by the CDM Executive Board (EB), which was set up by the United Nations Framework Convention on Climate Change (UNFCCC) in 2001. Before submitting their applications to the EB, project developers need to go through a series of complicated procedures and get permission from the Designated National Authority (DNA) in the host country first. CDM projects are externally verified and certified by "Designated Operational Entities" (DOEs), the third party organizations designated by the parties to the Kyoto Protocol.

Generally speaking, a typical China-based CDM project needs to go through the following stages before it can produce CERs.

Project Identification

Project owners and other project participants including international investors and potential CER buyers from Annex I countries discuss the feasibility, technology options, fund allocation, and other issues concerning the projects. A Project Idea Note (PIN) can be worked out with a series of factors including the additionality principle, transaction cost, project risk, and the way of funding being considered.

Project Design

After the initial contact and negotiations, project developers need to fill out the form of Project Design Document (PDD), which can be downloaded from the UNFCCC's official web site. The content of a PDD includes the description of the project, application of baseline methodology, crediting period of CERs produced by the project, calculation of greenhouse gas emission reductions, environmental impact, and opinions from other stakeholders. All projects must conform to China's sustainable development criteria.

Host Country's Approval

According to *Measures for Operation and Management of CDM Projects in China*,[18] the DNA in China is the National Development and Reform Commission (NDRC). Project initiators should submit their application including the PDD to the NDRC for approval. On the basis of the opinions from the National CDM Board, the NDRC approves or rejects the proposal. The Board is co-chaired by the NDRC and the Ministry of Science

---

[18]For full text of the regulation, please refer to "Measures for Operation and Management of Clean Development Mechanism Projects in China", Clean Development Mechanism in China (21, November 2005), http://cdm.ccchina.gov.cn/english/NewsInfo.asp?NewsId=905.

and Technology (MoST), and also includes members from the Ministry of Foreign Affairs (MoFA), the State Environmental Protection Administration (SEPA), China Meteorological Administration (CMA), and the Ministry of Agriculture (MoA). For each CDM application, it normally takes less than 2 months to receive a letter issued by the NDRC on the status of the approval.

Project Validation

After being approved by the DNA, project initiators should sign a contract with a qualified DOE for future validation. The DOE then starts to examine and appraise the project on the basis of the PDD and other CDM regulations. A validation report would be given by the DOE after the assessment.

Project Registration

If the project passes DOE's validation, the DOE will submit the validation report to the EB, applying for project registration. If the application passes the EB's examination, the project will be successfully registered with the United Nations, and relevant CERs can be produced by the project.

Project Implementation, Supervision, and Reporting

During the implementation process of the CDM project, project developers have the responsibility to report relevant data concerning emission, baseline, and leakage to the DOE.

Verification and Certification of CERs

The DOE examines the supervision report provided by project developers and verifies the emission reductions produced by the projects through various means including spot inspection. After verification, the DOE will submit a certification report to the project participants and the EB.

Issuance of CERs

After the EB approves certification application, 2% of the CERs will be deposited into a special account as climate change adaptation fund for least developing countries. The remaining CERs will go to the accounts of relevant project participants or Parties to the Kyoto Protocol according to the contract.

CHALLENGES AND OBSTACLES

China's move to further tap its CDM potential under the framework of Kyoto Protocol is not without its challenges. Many local corporations still lack the basic knowledge about KP and CDM, or the expertise to prepare Project Design Document (PDD). The complicated application procedures and considerable prophase investment are also obstacles for the launching of CDM projects, which needs the approval of both the Chinese government and the United Nations. Project initiators have to ensure that the CER price can cover their additional investment. Many projects concerning environmental protection and system upgrade need to use new equipment and technology; so, they sometimes become too risky for the commercial banks and other investors to participate in.

Most other developing countries, including many very poor ones that are extremely vulnerable to global warming, have been largely left out of the big pie. In 2006, all of Africa only received less than US$ 150 million from selling carbon credits, or 3% of the global total (Figure 10.1).

The imbalanced regional distribution in carbon trade income as well as China's surging energy consumption and emissions have led to international dissatisfaction about the current climate change regime dominated by the Kyoto Protocol. When the round of international talks kicked off at the end of 2007 to formulate a new pact replacing the Kyoto Protocol, increased international pressure was upon China to bear compulsory emission cutting. Environmental activists hope that, the China the current world's No.1 greenhouse gas emitter, plus United States and India, will all be included in the new treaty with fixed emission quotas or even carbon taxes on them. The United States, the notorious betrayer of the Kyoto Protocol, refused to accept any mandatory cuts in emissions with a similar commitment from the developing countries, notably China.

China is in a dilemma. On the one hand, it is fretted about its deteriorated environmental problem, with the hope to utilize international cooperation to solve it; on the other hand, it fears that emission abatement

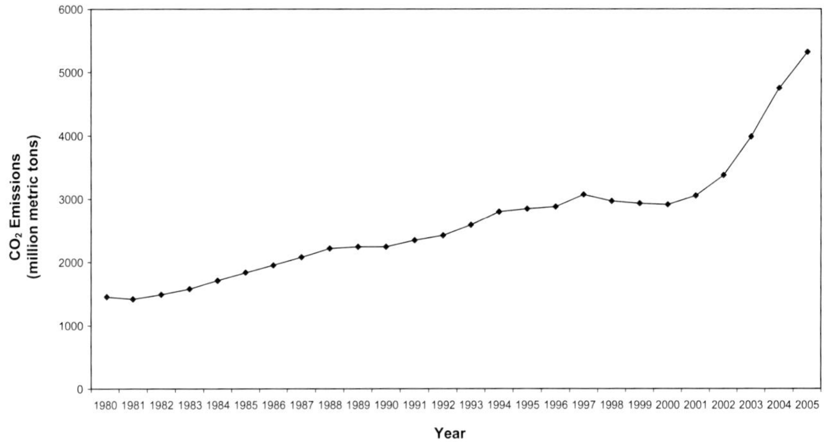

Figure 10.5. China's CO_2 Emissions from 1980–2005 (million metric tons).
Source: World Carbon Dioxide Emissions from the Consumption and Flaring of Fossil Fuels, US Energy Information Administration (EIA), http://www.eia.doe.gov/iea/carbon.html.

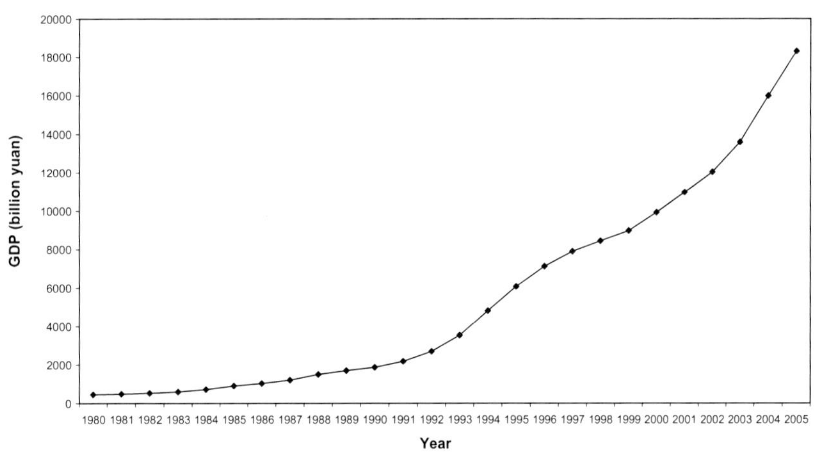

Figure 10.6. China's GDP from 1980–2005 (billion yuan).
Source: *GDP from 1978 to 2006*, China's National Bureau of Statistics, http://219.235.129.54/cx/table/table_sc.jsp?bh=0000000000000203&dzm=000000000&bbzl=101.

obligations may block its fast economic growth. China's total GDP is still less than one-fourth the size of the United States, but its total emissions have ballooned (Figures 10.5 and 10.6) because of an investment splurge on steel, cement, aluminum, and the like, all hugely energy-intensive industries, plus reliance on coal to produce power.[19] From China's perspective, international effort against climate change is a two-edged sword, one which can transfer fund and technology to China, as has already happened, but can also put shackles upon the giant if a disadvantage results from future negotiations.

Chinese experts argue that if China accepts emission quotas just like those upon industrialized nations in the Kyoto Protocol, it will finally pay much more than it currently gains from CDM projects to buy emission permits from other countries at a higher carbon price in the post-Kyoto period. The alternative carbon tax, if adopted in the next protocol, is expected to cost Chinese industry billions of dollars or more if China's economy and emissions continue to grow at the current rates.

[19]"Pressure Can Spur China into Climate Action", *Financial News* (9, May 2007).

Chapter 11

Reforms and New Policies

With mounting environmental problems across the country, the ruling Chinese Communist Party (CCP) included the notion of "Conservation Culture" (*shengtai wenming*) in its political report for the first time as a new content of "Building a Comprehensive Well-off Society" (*quanmian jianshe xiaokang shehui*). Environmental and resource cost to economic growth also topped all other problems and difficulties mentioned by CCP General Secretary Hu Jintao in his political report at the 17th Party Congress held in October 2007. Environmental issues, once marginalized in China's upmost decision-making process, have become new policy focuses and thus deserve more attention from China watchers.

NEW POLICIES AND INITIATIVES

Although the objective of building a comprehensive well-off society remains, the report of the 17th Party Congress has cited the "reduction of energy consumption and protection of the environment" alongside the last report's "optimization of structure and improvement of economic results" as the bases for economic development. This change means that "coordinated development" with the focus on green issues will permeate future economic reform. If the CCP wants to consolidate its rule, it has to address environmental issues properly and timely.

In fact, new stringent policies and programs, economic incentives, and disincentives as well as institutional changes are already under way to punish those violating environmental rules and to encourage more environmentally-friendly actions.

Suspension of Approval of New Industrial Projects along Major River Systems (*liuyu xianpi*)

On 3, July 2007, the SEPA announced that no new industrial projects would be approved in several cities and industrial parks along four major river systems to prevent them from being further contaminated. Six cities, two counties, and five industrial zones were indicted by the SEPA for their role in polluting the Yangtze, Yellow, Huaihe, and Haihe rivers. The cities are Chaohu and Bengbu in Anhui, Baiyin in Ningxia, Bayannur in Inner Mongolia, Weinan in Shaanxi, Zhoukou in Henan; and the two counties are Hejin and Xiangfen in Shanxi. The industrial parks are in Wuhu in Anhui, Lanzhou in Gansu, Handan in Hebei, Puyang in Henan and Shenxian County in Shandong. The SEPA will not approve any projects for 3 months apart from treatment plants and recycling facilities, and the ban will not be lifted until the sources of untreated wastewater are shut down and treatment facilities installed. This kind of suspension of approval of new industrial projects in heavily polluted areas along river systems (*liu yu xian pi*) is regarded by SEPA as the toughest measure to fight against water pollution, given its limited authority.

Electricity Generated by Renewable Energy to Be Purchased Preferentially

From September 2007, China's State Electricity Regulatory Commission (SERC) started acquiring all power grid companies in China to preferentially purchase the maximum amount of "green" electricity generated from renewable sources such as hydropower, wind power, biomass, solar power, tidal power, and geothermal energy, according to a new regulation released by SERC.

The regulation details the authority, measures, and responsibilities necessary for SERC to facilitate the integration of renewable sources into power systems. It allows all renewable power facilities, with the exception of medium- and large-scale hydropower plants, to receive government subsidies in power pricing rather than having to participate in competitive bidding.

China adopted its first law on renewable energy in 2005 and has since issued several supplementary rules and regulations to boost the use of renewable energy. Yet, renewable sources still account for only a very small portion of domestic power supply, mainly because of the high generation costs. The cost of electricity generated from solar power, for example, is some 3 yuan (40 US cents) per kilowatt-hour, while that from a typical coal-fired power plant is only around 0.22 yuan (2.8 US cents) per kilowatt-hour.[1]

A supplementary regulation on renewable power pricing and cost sharing, authored by the National Development and Reform Committee, has helped break this cost bottleneck by requiring power suppliers on the grid to purchase renewable electricity at either a government-fixed or a government-directed price. The additional cost of renewable energy is to be borne by electricity users. An extra "renewable energy" charge of 0.001 yuan (0.013 US cents) for every unit of electricity has been added to household utility bills since June 2006.

"Green-Credit Policy" (*luse xindai zhengce*)

In July 2007, SEPA, the People's Bank of China (PBC) and the China Banking Regulatory Commission (CBRC) jointly issued a "green-credit policy", which required SEPA to hand over lists of heavy polluters to the central bank and the regulatory commission. The companies on the list that fail to pass environmental assessments or to implement China's environmental protection regulations are disqualified from receiving loans from any bank or financial institution. The policy is regarded as the first one of its kind in the country.[2]

---

[1] Ling Li, "China Urges Electricity Suppliers to Buy 'Green' Power", World Watch Institute (30, August 2007), http://www.worldwatch.org/node/5330.

[2] Chen Xia, "New Policies to Reinforce Environmental Protection", China Internet Information Center (12, September 2007), http://english.china.org.cn/english/environment/224226.htm.

Companies that already have loans, but are later discovered to have violated environmental protection regulations, will have their loans recalled, according to the policy. The SEPA has said it would join with the Ministry of Finance, the China Insurance Regulatory Commission, and the China Securities Regulatory Commission to devise other economic policies to promote environmental protection. By November 2007, 12 heavy polluting enterprises had had crucial bank loans recalled, suspended, or rejected as China's new "green-credit policy" kicks into action. The 12 companies were among the 30 the SEPA reported to the People's Bank of China and the China Banking Regulatory Commission since July 2007.[3]

Establishing Leading Group on Energy Saving and Emission Cutting

Pressured by the international community for its rapidly-growing carbon emissions, China in 2007 established the National Leading Group to Address Energy Saving, Emission Cutting and Climate Change, headed by Premier Wen Jiabao.

The country decided to set up such a high-level leading group to coordinate the inter-agency relations concerning industrial production and environmental protection. China's long-time "development first" stance is closely related to its special inter-agency mechanism on energy saving and climate change, in which the National Development and Reform Commission (NDRC) plays a dominant role while the State Environmental Protection Administration (SEPA) and China Meteorological Administration (CMA) playing minor roles.

The newly-established National Leading Group on Energy Saving and Emission Cutting is expected to emphasize more on the environment aspect in the whole policy-making process, but the NDRC's strong position will not be changed much as the National Leading Group puts its regular office inside the NDRC.

---

[3]"12 Heavy Polluters Punished Under New 'Green-Credit' Policy", *Xinhua News* (16, November 2007).

Using Green Indexes to Measure Officials' Performances

On 23, November 2007, the State Council released a circular that uses green figures of per unit GDP energy consumption and major pollutants reduction to assess provincial officials' performances. Since early 1990s, China's officialdom has formed a culture that GDP growth is a major indicator to measure administrative chiefs' performances and even determines their future promotions. Meanwhile, local officials are incentivized to promote pure economic growth, as it is the source of the local fiscal revenue increase.

Instead of pure economic figures, the Chinese government is trying to measure local officials' achievement through a much more complicated system that also includes environmental indexes. According to the "Implementation Plan for Unit GDP Energy Consumption Assessment System" approved by the State Council and included in the circular, major officials of the provincial-level regions as well as leaders of the top-1000 state-owned or controlled energy consumption enterprises are to be assessed by per unit GDP energy consumption figures calculated through quantified scoring methodology.

For all provincial governments, the energy conservation assessment results will be provided to the personnel department, and will be regarded as an important basis for the comprehensive assessment of the leading group and leading officials of the government. The accountability system and one ballot veto system will be implemented. For those provincial governments which are evaluated as failing to meet those targets, the leading officials shall not be eligible to take part in the annual award competition, or to receive any honorary title. The central government will temporarily halt the verification and approval of newly constructed high energy consuming projects in this region. For those top-1000 key enterprises which are evaluated as failing to meet the targets, they will be subject to a notice of criticism, and shall not be allowed to attend the annual award competition, or to receive any honorary title or be granted with inspection-free and other incentive treatments. The verification and approval for their newly constructed high energy consuming investment projects and newly added industrial land, will be suspended.

The Supervision and Management Agency for State Owned Assets will take the assessment results for sole state-invested or state holding enterprises as an important basis for the performance assessment of the leaders of those top-1000 key enterprises. The one ballot veto will be implemented. For those provincial governments and top-1000 key enterprises which were evaluated as failing to meet the targets, they shall be asked to take actions for rectification and upgrading during a given time. For those who fulfilled or over-fulfilled the assessment targets, commendation and awards will be granted.

Establishment of Green Fund

In November 2007, China launched a state-owned Clean Development Mechanism (CDM) fund to finance the country's efforts to address climate change.[4] The China CDM Fund, managed by the Ministry of Finance, generates money from the current CDM projects that help China improve energy efficiency and protection of the environment by using clean energy for power generation. By the end of October, the NDRC, China's top economic planner, had approved 885 CDM projects that were expected to supply 1.5 billion metric tons of carbon credits, said Xie Zhenhua, NDRC deputy director. He added that the projects would generate 15 billion US dollars, of which three billion US dollars would be injected into the fund.[5] The fund also receives donations from financial institutions such as the World Bank and the Asian Development Bank.

Eco-City Project

China wants to improve its international image on environmental protection through starting bilateral or multilateral green projects, with the landmark China–Singapore "Eco-City" Project being the most high-profile one.

During Singaporean Senior Minister Goh Chok Tong's meeting with Chinese Premier Wen Jiabao in Beijing in April 2007, Goh firstly proposed jointly developing with Beijing an "Eco-City" in China. This coincided with Wen's strong demand for much tougher action domestically to deal

---

[4] "China Launches CDM Fund to Address Climate Change", GOV.cn (9, November 2007), http://english.gov.cn/2007-11/09/content_801221.htm.
[5] *Ibid.*

with the grave environmental problems, winning Beijing's interest and endorsement in principle. During Chinese Vice Premier Wu Yi's recent visit to Singapore in July 2007, the Eco-City plan was again emphasized during the bilateral talks and it was expected to serve as an iconic project for sustainable development that the Chinese leadership is very eager to develop.

During Chinese Premier Wen Jiabao's visit to Singapore in November 2007, he and Singaporean Prime Minister Lee Hsien Loong finalized this flagship Eco-City project in Tianjin through signing a framework agreement. The Eco-City would be built in the Binhai New Area of Tianjin Municipality, a state-level strategic development area approved by the State Council in 2005 for its huge potential for economic growth. The eco-city would comprise a 30-square kilometer "town", combining modern living, recreational, and business facilities. Tianjin and Singapore would form a joint venture for the project, participated by firms from both countries, with Singapore's Keppel Corporation playing a lead role. The project is expected to use advanced energy-saving and environmental technologies in the hopes of turning Tianjin into a model "green" city. Singapore government officials said the project's management, technology, and policy aspects will serve as a reference for other Chinese cities.

Discussion on Levy of Environmental Tax

The levying of an environmental tax is on the Chinese government's agenda in 2008 to promote environmental protection. The environment tax, one of the key taxation policies that China plans to push forward in 2008, is going through careful study and appraisal by the Ministry of Finance, State Administration of Taxation, and SEPA.[6]

Two researchers with the Ministry of Finance, Sun Gang and Xu Wen, proposed three plans for the collection of environment tax in their report published in November 2007. Based on the principle of "pay according to profits", the first plan suggests that the government collect tax from companies in accordance with the amount they earn from their products

[6]"Government Mulls Tax Plan for Environmental Protection", *Xinhua News* (7, January 2008).

that consume resources or pollute the environment, said Xu. He said that this kind of general taxation will be imposed on all the companies since almost all the products consume resources. According to the second plan, companies would pay tax according to how much pollutant they have discharged, said Xu, adding that the tax will target the discharge of sulfur dioxide, carbon dioxide, sewage, solid waste, and other forms of pollutant. He said that the third plan would ask consumers of potentially polluting products, such as fossil fuel, ozone-consuming commodities, fertilizer, and pesticide, to pay the environmental tax.[7]

Meanwhile, the Ministry of Finance said it might be a more pressing task for them to raise the resource and consumption tax, to give more consideration to environmental protection while expanding the current taxation system and adding new kinds of tax.

Establishing Pilot Zone on Resource Efficiency and Environmental Protection

At the end of 2007, the cities surrounding Changsha and Wuhan in the middle reaches of the Yangtze River decided to ally to form a new pilot zone with the aim of building a resource-efficient and environment-friendly society. This became China's fourth pilot zone for comprehensive and coordinated economic reform. The first three state-level pilot zones are Shanghai Pudong, Tianjin Binhai, and Chongqing-Chengdu pilot zone. For the Wuhan and Changsha clusters, the focus will be on sustainable growth, efficiency, and environmentally friendly projects.

The pilot zone will include two clusters of cities in Central China: one consists of Wuhan, capital of Hubei Province, and eight surrounding cities in Hubei, and the other is composed of Changsha, Zhuzhou, and Xiangtan, the three major cities in Hunan Province. The task will involve systematic planning of reforms in production, consumption, circulation, and other fields, choosing between different strategies of development as well as launching institutional reforms.[8]

---

[7] *Ibid.*

[8] Tan Wei, "Hope for Central China", *Beijing Review*, No. 49 (2007), http://www.bjreview.com.cn/business/txt/2007-12/01/content_88201.htm.

The two clusters are labeled as bases for modern equipment manufacturing and hi-tech industries. The Wuhan cluster is the largest distribution center of automobiles and auto parts and is building a high-tech base dominated by photoelectronic information technologies.[9] The Changsha cluster possesses many of the top innovations in China, such as walking robots, the Queen Wasp series of remote-controlled mini aircraft, and the Yinhe Super-Simulation Computers.[10] If the experiment on these two traditional industrial bases is successful, it will set good examples for other parts of China to fulfill sustainable development.

Promoting "Green Insurance"

After the introduction of green credits to use market instruments to treat environmental problems, the country's top environmental watchdog and finance regulator took a second step through jointly promoting insurance for polluters and help expedite compensation for victims. The latest pollution liability insurance policy, to be put in place in 2015, covers the clean up of environmental accidents and indemnifies the insured from pollution treatment costs.[11]

According to the new policy of "green insurance", insurance products will be promoted to three types of industrial plants with high risks of environmental pollution mishaps — hazardous chemical plants involved in production, selling, storage and transportation, petrochemical plants, and hazardous waste treatment plants. "Without pollution liability insurance, the culprits could not compensate victims and conduct cleanups, generating social instability", SEPA Deputy Director Pan Yue said.[12] With this insurance scheme, the government is able to alleviate financial burden from serious pollution incidents, and the polluter (usually industrial plants) can avoid going bankrupt with indemnities while residents can also get compensated for suffering the pollution. Due to the high costs involved, however, most enterprises were not active in buying such green

[9] *Ibid.*
[10] *Ibid.*
[11] Sun Xiaohua, "Polluters Urged to Buy Green Insurance", *China Daily* (19, February 2008).
[12] *Ibid.*

insurance, according to a survey by SEPA and China Insurance Regulatory Commission.[13] A possible way for the SEPA to solve the problem is to list the purchase of such insurance as prerequisites for proving the environmental impact assessment of projects or in allocating pollution discharge licenses.

Elevating SEPA to Full-Fledged Ministry — Ministry of Environmental Protection

The Chinese government elevated the SEPA to a fully fledged environment ministry during the annual session of the 11th National People's Congress in March 2008. This was a noticeable part of the government's ambitious "super-ministries" reform (*dabuzhi gaige*), which aimed to merge ministries and commissions with similar functions and overlapping powers into larger-sized ministries. The creation of the Ministry of Environmental Protection gives its representatives the power to influence cabinet-level decisions on green issues, helping them to join important meetings of the State Council and ensuring environmental protection efforts are better integrated into the decision-making process. Equipped by more staff and greater financial support, the new Ministry of Environmental Protection will be responsible for drafting and implementing programs, policies, and standards concerning environmental protection; working out environmental functions in different regions; supervising pollution prevention and treatment efforts; and tackling major environmental issues.

Compared with SEPA, its predecessor, the Ministry of Environmental Protection is expected to have a consolidated authority over pollution control, which was previously handled by several different departments. Although the Ministry of Water Resources still retains the power of protecting water resources, the Ministry of Environmental Protection now takes full responsibility for water pollution prevention and water quality monitoring. It has been clarified that the Ministry of Environmental Protection (not the Ministry of Water Resources) has the authority to release information about water environment. The Ministry of Water Resources and SEPA had been wrangling over this issue for years.

---

[13] *Ibid.*

The Ministry of Environmental Protection has expanded moderately after its establishment, with three new departments set up, namely the Department for Control of Pollutant Discharge, the Environment Monitoring Department, and the Department of Publicity and Education. The move highlighted the ministry's duties to prevent pollution at its source, react to environmental incidents, and promote public awareness about environmental protection, showing the central government's greater concern about the environmental issues. Local environmental protection bureaus' (EPBs) functions and range of power change correspondingly, but their high dependence upon local party and administrative chiefs remain unchanged.

A LONG WAY TO GO GREEN

The series of stringent measures taken by the Chinese government to tackle the environmental problems have shown some limited effect. In 2007, some leading indicators of environmental degradation turned slightly better. In year 2007, China's two major pollution indicators, SO_2 and chemical oxygen demand (COD), dropped for the first time by 4.66% and 3.14%, respectively year on year, and per-unit GDP energy consumption decreased by 3.27%.

These figures, however, were still far from meeting the requirement set by China's 11th Five-Year Plan (2006–2010). Under this plan, the Chinese government set forth a number of ambitious green targets, including cutting energy consumption per unit GDP by 20%, reducing major pollutants discharge by 10%, decreasing water consumption per unit industrial output by 30%, and raising forestry coverage rate to 20%.

In 2006, the country only managed to cut its energy intensity by 1.23%, far below the annual 4% target in the 11th Five-Year Plan. Reports from the SEPA showed that in the first quarter of 2007, China's seven major water systems were slightly polluted, with no major change in water quality from the same period of 2006. The overall water quality of drinking water sources dropped, with only 69.3% of drinking water sources reaching national standards, down 5% year on year. Prior to the 2008 Olympics, air quality in Beijing had not become any better, which embarrassed the government and forced officials to expand harsh emission control measures to neighboring areas at the eve of the Games.

The environmental crisis in China, probably unmatched for its sheer scale and for the speed with which it is unfolding, increasingly attracts the attention of both sinologists who see the significance of these environmental challenges for wider Chinese politics, and environmental scholars who recognize the importance of China in global environmental politics.[14] People have to admit that environmental pollution is to some extent an inevitable by-product of rapid industrialization and urbanization process of such a large-size state, and that China has achieved remarkable improvements in its environmental governance since late 1970s. People have found some similarities between the economy–pollution correlation in China and that of many industrialized nations in their early taking-off days. What is worrying, however, is the evident distinction of China's special political systems that in many areas hamper it from taking more effective actions to handle the urgent environmental problems. For many western countries, environmental issues were novelties that gradually drew the attention from the government and the public while on their way to modernization and industrialization. As a latecomer in industrialization, China had learnt a lot from the west and accumulated plenty of past experiences and relevant knowledge about the seriousness of environmental pollution for a long period of time.

Despite such intensive and long-lasting attention, the government still feels frustrated in fighting against ecological degradation, which is worth concern and further deliberation. Sufficient funding and green technology have no longer become major obstacles for China to fulfill green targets. With neck-breaking economic growth of yearly 10% for the past three decades, China's intrinsic financial and technological capability have been greatly enhanced during the process of building up the market economy. Meanwhile, China has proved itself to be highly efficient in attracting foreign capital and technology in various sectors. We should, therefore, go to the political realm to find out the structural roots of the

[14]Neil T. Carter and Arthur P.J. Mol, "China and the Environment: Domestic and Transnational Dynamics of a Future Hegemon", in Neil T. Carter and Arthur P.J. Mol (eds.), *Environmental Governance in China* (London: Routledge, 2007), p. 183.

quickly-deteriorating environment and the potential the country owns in environmental governance after significant institutional and policy changes. To some extent, China is indeed catching up with more developed countries in designing and implementing a more advanced system of environmental governance to meet current and future challenges, but by the same token China is still different, and its environmental governance system carries the consequences of the national particularities of a transitional state with a different cultural background.[15]

The Taihu algae outbreak showed that the government's priority on economic development for legitimization of the regime could be the fundamental factor leading to frequent environmental disasters in China, while lack of supervision or an active civic society would exacerbate the situation. The toxic spill crisis in the Songhua River reveals on the one hand that the current authoritarian political system possesses efficient administrative capability to swiftly diminish the negative impact of serious environmental crisis in a short time after it is known to the public, while on the other hand, information opaqueness and officials' deliberate cover-up often make the regime miss the best opportunity and timing to take decisive actions and prevent hidden troubles from evolving into catastrophes. China's participation into the global collective action against climate change tells that the country could establish one of the world's most efficient and systematic management structure to handle burgeoning green business, but if the green target is expected to decelerate economic growth, the government will be especially reluctant to shoulder relevant international responsibilities.

Chinese leaders have to enhance the environmental protection sectors to solve the pollution problem more effectively. They should introduce economic incentives and disincentives to curb pollution and ecological destruction instead of relying too much on short-term administrative orders such as closing polluting enterprises or blocking pollution sources. Institutional innovations such as the levy of reasonable discharge fees as well as resource, fuel and environmental taxes can do a better job than administrative orders as they are fair and have long-lasting effect.

[15]*Ibid.*, p. 185.

In fact, China has already made impressive effort to promote market-based tradable permits for SO_2 emissions control. As the world's largest SO_2 emitter, China suffers greatly from the negative impacts of acid rain on agricultural production, respiratory health, and the longevity of buildings. Through learning from mature emission trading experience from the United States and other western countries, China initiated a Total Emissions Control (TEC) policy, in which the government annually set caps for the total SO_2 emissions throughout the country, and divided the quota among the provinces, autonomous regions, and municipalities, who were then responsible for assigning their own TEC targets to local governments or emission contributors.[16] Within the overall policy of TEC, China began conducting trials in emissions trading, working with a number of international partners, including Resources for the Future, the Asian Development Bank, the US Environmental Protection Agency, and Environmental Defense.[17] It is definitely a brave and innovative attempt in a correct direction, but certain weak points in the country's overall governance system, such as lack of supervision and information transparency, weak environmental enforcement capacity, and disunity in emission methodologies among different regions, restrained such market-based scheme from playing an effective role in controlling emissions. Without the significant improvement of China's institutional and regulatory environment as a whole, it is difficult for such market-based incentives to produce the same impact in China as in western countries.

Increased public participation is also essential for environmental protection. In China's state-centric society, the forces of environmental NGOs are still very weak and the media's supervision is quite limited. Political and sociocultural dynamics influence the opportunities for citizen consumers to become involved in the management of the environmental commons, and the involvement of citizen consumers in environmental

---

[16]For details of the SO_2 emission trade scheme, please refer to Wang, J., Yang, J., Ge, C., Cao, D. and Schreifels, J., "Controlling Sulfur Dioxide in China: Will Emission Trading Work?" *Environment*, Vol. 46, No. 5 (2004), pp. 28–39.

[17]Elizabeth Economy, "Environmental Governance: The Emerging Economic Dimension", in Neil T. Carter and Arthur P.J. Mol (eds.) *Environmental Governance in China* (London: Routledge, 2007), p. 36.

reform is also related to the environmental conditions in a particular locality, including the biophysical state of the environment, the level of public environmental awareness, and the environmental management situation.[18]

The concept of Green Olympics helped many residents in Beijing and elsewhere in China to change their habits to embrace a greener lifestyle. There has been a growing awareness among the public of energy saving and environmental protection after the Olympics. Some Beijing residents quit driving their cars after Beijing adopted the plan of allowing cars on the road on alternate days based on license plates ending in odd or even numbers. During the Games, even on the days when their cars are allowed to be on the road, some people still rode bicycles or took public transportation to and from work. Such subtle changes in daily lifestyles also include shifting from the use of air conditioners to electric fans in the summer, purchase of energy-saving light bulbs, and dumping trash in sorted bins.

In China, increasing media coverage on the threat of climate change and the urgency of environmental protection has contributed to raising public awareness. Many environmental organizations have produced creative media campaigns to generate public interest, including for the Olympics. Teaming up with 11 former Olympic gold medalists, the China Environment Awareness Program (CEAP) created eight public service advertisements to raise public environmental consciousness toward a "Green Olympics." Although each of the public service ads lasted only 30 seconds, the eight advertisements address key environmental protection issues, such as sorting trash for recycling, saving water resources, protecting biodiversity, and using more clean energy. These public service ads were broadcast on various TV channels, on the monitor screens within the Olympic facilities and stadiums and on public transportation systems.

The CCP's new concept of "Conservation Culture", therefore, only marked the beginning of the long-term effort to clean up the country's severely polluted environment. The world's most populous country certainly still has a long way to go before the balance between economy and ecological systems can be regained.

---

[18]Susan Martens, 'Public Participation with Chinese Characteristics: Citizen Consumers in China's Environmental Management', in Neil T. Carter and Arthur P.J. Mol (eds.) *Environmental Governance in China* (London: Routledge, 2007), pp. 66–67.

Bibliography

BOOKS AND BOOK CHAPTERS

Abigail R. Jahiel, "The Organization of Environmental Protection in China", in Edmonds, Richard Louis (ed.), *Managing the Chinese Environment* (Oxford: Oxford University Press, 1998).

Carlos Wing-Hung Lo and Shui-Yan Tang, "Institutional Reform, Economic Changes, and Local Environmental Management in China: The Case of Guangdong Province", in Neil T. Carter and Arthur P.J. Mol (eds.), *Environmental Governance in China* (London: Routledge, 2007).

China Environment Yearbook Editing Committee (ed.), *China Environment Yearbook (1990)* (Beijing: China Environmental Science Press, 1990).

China Environment Yearbook Editing Committee (ed.), *China Environment Yearbook (1996)* (Beijing: China Environmental Science Press, 1996).

China Environment Yearbook Editing Committee (ed.), *China Environment Yearbook (2001)* (Beijing: China Environmental Science Press, 2001).

China Environment Yearbook Editing Committee (ed.), *China Environment Yearbook* (2006) (Beijing: China Environmental Science Press, 2006).

Dasgupta, C., "The Climate Change Negotiations", in I. Mintzer and J. Leonard (eds.), *Negotiating Climate Change: The Inside Story of the Rio Convention*, (Cambridge: Cambridge University Press, 1994).

Elizabeth Economy, "Environmental Governance: The Emerging Economic Dimension", in Neil T. Carter and Arthur P.J. Mol (eds.), *Environmental Governance in China* (London: Routledge, 2007).

Fu Tao, "Development of Environmental NGOs in China", in Liang Congjie and Yang Dongping (eds.), *The China Environment Yearbook 2005: Crisis and Breakthrough of China's Environment* (Boston: Brill, 2007).

Jin Jiaman, "The Growing Importance of Public Participation in China's Environmental Movement", in Jennifer Turner and Fengshi Wu (eds.), *Green NGO and Environmental Journalist Forum: A Meeting of Environmentalists in Mainland China, Hong Kong, and Taiwan* (Washington, DC: The Woodrow Wilson Center, 2001).

Lester Ross, "China: Environmental Protection, Domestic Policy Trends, Patterns of Participation in Regimes and Compliance with International Norms", in Richard Louis Edmonds (ed.), *Managing the Chinese Environment* (Oxford: Oxford University Press, 1998).

Liang Congjie and Yang Dongping (eds.), *The China Environment Yearbook 2005: Crisis and Breakthrough of China's Environment* (Boston: Brill, 2007).

Lu Xuedu and Liu Deshun (eds.), *Clean Development Mechanism in China: Taking a Proactive and Sustainable Approach* (Mandarin Edition), (Beijing: Tsinghua University Publishing House, 2004).

Ma Jun, *China's Water Crisis* (Norwalk: EastBridge, 2004).

Neil T. Carter and Arthur P.J. Mol (eds.), *Environmental Governance in China* (London: Routledge, 2007).

Nye, Joseph S. Jr., *Soft Power: The Means to Success in World Politics* (New York: Public Affairs, 2004).

Office of the National Coordination Committee on Climate Change and Institute of Nuclear and New Energy Technology of Tsinghua University, *Guidelines on China's CDM Project Development* (Beijing: China Environmental Science Publishing Company, 2006).

Peter M. Haas and Jan Sundgren, "Evolving International Environmental Law: Changing Practices of National Sovereignty", in Nazli Choucri (ed.), *Global Accord: Environmental Challenges and International Responses* (Cambridge, MA: The MIT Press, 1993).

Qie Jianrong, "Warnings and Lessons from the Songhua River Toxic Spill Crisis", in Liang Congjie and Yang Dongping (eds.), *The China Environment Yearbook 2005: Crisis and Breakthrough of China's Environment* (Boston: Brill, 2007).

Richard Louis Edmonds (ed.), *Managing the Chinese Environment* (Oxford: Oxford University Press, 1998).

Susan Martens, "Public Participation with Chinese Characteristics: Citizen Consumers in China's Environmental Management", in Neil T. Carter and Arthur P.J. Mol (eds.), *Environmental Governance in China* (London: Routledge, 2007).

Vaclav, Smil, *China's Past, China's Future: Energy, Food Environment* (New York: RoutledgeCurzon, 2004).

Xiaoying Ma and Leonard Ortolano, *Environmental Regulation in China: Institutions, Enforcement, and Compliance* (Oxford: Rowman & Littlefield Publishers, 2000).

Yuka Kobayashi, "Navigating Between 'Luxury' and 'Survival' Emissions", in Paul G. Harris (ed.), *Global Warming and East Asia* (London: Routledge, 2003).

Zhang Zhihong, "The Forces Behind China's Climate Change Policy", in Paul G. Harris (ed.), *Global Warming and East Asia* (London: Routledge, 2003).

JOURNAL ARTICLES

Carmen Richerzhagen and Imme Scholz, "China's Capacities for Mitigating Climate Change" *World Development*, Vol. 36, No. 2 (2007).

David G. Street, *et al.*, "Air Quality During the 2008 Beijing Olympic Games", *Atmospheric Environment*, No. 41 (2007).

Eduard B. Vermeer, "Industrial Pollution in China and Remedial Politics", *The China Quarterly*, No. 156 (1998).

Elizabeth C. Economy, "The Great Leap Backward?" *Foreign Affairs*, Vol. 86, Issue 5 (2007).

Hong Liu, "Profit or Ideology? The Chinese Press between Party and Market", *Media, Culture and Society*, No. 20 (1998).

J. Wang, J. Yang, C. Ge, D. Cao and J. Schreifels, "Controlling Sulfur Dioxide in China: Will Emission Trading Work?", *Environment*, Vol. 46, No. 5 (2004).

Jianguo Liu and Jared Diamond, "Revolutionizing China's Environmental Protection", *Science*, Vol. 319, No. 5859 (2008).

Joanna Lewis, "China's Strategic Priorities in International Climate Change Negotiations" *The Washington Quarterly*, Vol. 31, No. 1 (2007).

Kevin J. O'Brien, "Neither Transgressive Nor Contained: Boundary-Spanning Contention in China", *Mobilization*, Vol. 8, No. 1 (2003).

Robert O. Keohane, "International Institutions: Can Interdependence Work?", *Foreign Policy*, Issue 110 (1998).

Stefanie Beyer, "Environmental Law and Policy in the People's Republic of China", *China Journal of International Law*, Vol. 5, No. 1 (2006).

Wen Bo, "Greening the Chinese Media", *China Environment Series*, Issue 2 (1998).

Yang Guobin and Craig Calhoun, "Media, Civil Society, and the Rise of a Green Public Sphere in China", *China Information*, No. 21 (2007).

Yang Guobin, "Environmental NGOs and Institutional Dynamics in China", *The China Quarterly*, No. 181 (2005).

NEWS MEDIA

AP (Associated Press) News
Beijing Review
Bloomberg News
China Business News
China Daily
Financial News
International Herald Tribune
People's Daily
Shanghai Securities News
South China Morning Post
Southern Metropolitan News
Southern Weekly
The Economist
The New York Times
The Wall Street Journal
Xinhua News

OTHER REFERENCES

Cui Dapeng, Analysis of International Negotiation on Climate Change from the Perspective of Political Economy. Doctoral Dissertation, Beijing: Renmin University of China (June 2002).

National Development and Reform Commission (NDRC), *China's National Climate Change Programme* (2007).

OECD Report, *Action Against Climate Change: The Kyoto Protocol and Beyond* (1999).

OECD Working Party on Environmental Performance, *Environmental Performance Review of China — Conclusions and Recommendations* (Final) (2006).

State Environmental Protection Administration (SEPA), *Gazette of China's Environmental Situation* (2005).

World Bank and SEPA's Joint Report, *Cost of Pollution in China: Economic Estimates of Physical Damages* (2007).

World Bank Report, *The Effects on Developing Countries of the Kyoto Protocol and CO_2 Emission Trading* (1998).

WEBSITES

Beijing Environmental Protection Bureau, http://www.bjepb.gov.cn.

China's Clean Development Mechanism (CDM), http://www.ccchina.gov.cn.

China's Government, http://www.gov.cn.

China's Ministry of Environmental Protection, http://english.mep.gov.cn.

Intergovernmental Panel on Climate Change (IPCC), http://www.ipcc.ch.

International Union for the Conservation of Nature and Natural Resources (IUCN), http://cms.iucn.org.

Jiangsu Provincial Environmental Protection Bureau, http://www.jshb.gov.cn.

National Bureau of Statistics (China), http://www.stats.gov.cn.

Netherlands Environmental Assessment Agency, http://www.mnp.nl/en.

Organisation for Economic Co-operation and Development (OECD), http://www.oecd.org.

Shanghai Population and Development Research Center, http://www.spic.sh.cn.

State Environmental Protection Administration (SEPA), http://www.sepa.gov.cn.

United Nations Framework Convention on Climate Change (UNFCCC), http://unfccc.int.

World Bank, http://www.worldbank.org.

World Wildlife Fund (WWF) China, http://www.wwfchina.org.

Index

10th Five-Year Plan, 14, 77
11th Five-Year Plan, 13, 78, 113, 154
17th Party Congress, 144
21st century, 1
3Rs (Reduce, Reuse, Recycle), 13
6th Five-Year Plan, 57

Abigail R. Jahiel, 19, 22, 29
accountability, 24, 74, 148
acid rain, 51, 157
additionality, 135, 139
afforestation, 91, 98, 126
Africa, 124, 141
Agenda 21, 106
Agreement on Trade Related Aspects of Intellectual Property Rights (TRIPS), 64
air pollution, 4, 8–10, 31, 34, 67, 89–92, 94, 97, 100, 102, 106
air pollution index, 91
air pollution law, 106
algae outbreak, 6, 76, 81, 156
All-China Environment Federation, 42
anarchical, 63, 110
aniline, 67
Annex I, 104, 135, 139
Argentina, 124
Arthur P. J. Mol, 24, 29, 104, 155, 157, 158
Asia-Pacific Economic Cooperation (APEC), 104, 125, 126
Asia-Pacific Partnership on Clean Development and Climate, 125
Asian Development Bank, 120, 149, 157
Association of Southeast Asian Nations (ASEAN), 125
Australia, 90, 125
Austria, 137

Baiyin, 145
Bali, 12, 121, 122, 124
Bangkok, 58
Basel Convention, 58
Bayannur, 145
Beijing, 3, 7–12, 43–45, 49, 56, 58, 59, 71, 73, 82–84, 88–102, 113, 116, 119, 120, 123, 128, 130, 135–137, 149–151, 154, 158
Beijing Environmental Protection Bureau, 89, 101
Bengbu, 145
benzene, 6, 67–69
Berlin, 58
Berne Convention, 64
Binhai New Area, 120, 150
biodiversity, 14, 15, 51, 56, 158
biology, 50
biomass, 136, 145
biosphere, 14
blogs, 48
blue sky day, 91–94, 100
Bohai Bay, 4
Bohai Sea, 4
Bonn Convention, 15
Brazil, 7, 60, 124
Britain, 27, 90, 121, 137
bulletin boards, 48
Bureau of Forestry, 21

Cambodia, 32
Canada, 7, 57, 59, 115, 128, 136, 137

cancer, 6
carbon dioxide (CO_2), 10–12, 29, 62, 98, 104, 114, 117, 130–133, 135, 142, 151
carbon finance, 135
carbon market, 123, 131, 135, 136
carbon monoxide, 22, 89, 95, 101, 118
carbon neutral, 98
carbon tax, 121, 122, 141, 143
central bank, 21, 146
certified emission reductions (CERs), 111, 128, 130–132, 134, 135, 137–141
Changchun, 72
Changsha, 56, 151, 152
Changshu, 131
Changzhou, 79
Chaohu, 2, 6, 76, 145
chemical oxygen demand (COD), 7, 79, 96, 154
Chen Jinsong, 70
Chen Qingfeng, 81
Chen Zhili, 99
Chengdu, 56, 151
China Association of the Environmental Protection Industry, 43
China Council for International Cooperation on Environment and Development (CCICED), 57
China Energy Conservation Association, 43
China Environment Awareness Program (CEAP), 158
China Environmental Culture Promotion Association (CECPA), 44
China Insurance Regulatory Commission, 147, 153
China Meteorological Administration (CMA), 20–22, 117, 118, 140, 147
China National Petroleum Corporation (CNPC), 6, 71, 72, 74
China Securities Regulatory Commission, 147
China Society of Environmental Sciences, 43
Chinese Communist Party (CCP), 16, 17, 49, 144, 158
Chinese People's Political Consultative Conference (CPPCC), 83
Chinese Society for Sustainable Development, 43
Chongqing, 3, 71, 75, 137, 151
circular economy, 13
civil society, 41, 47
Clean Development Mechanism (CDM), 29, 107, 108, 110, 111, 113, 117, 120, 122, 124, 127–141, 143, 149
clean water, 4, 79
climate change, 9–13, 21, 22, 50, 51, 55, 56, 58, 60, 62, 64, 86, 96, 98, 103–109, 114–130, 133–138, 140, 141, 143, 147, 149, 156, 158
Climate Exchange, 123
coal, 9–12, 31, 88, 94, 98, 112, 113, 116, 117, 137, 143, 146
Colby Pearce, 90
cold war, 55
Committee on Natural Resources and Environmental Protection, 19
common but differentiated responsibilities, 106, 108, 122
common-pool resources, 63
comparative advantage, 30
Comprehensive Nuclear-Test-Ban Treaty, 64
compressed natural gas, 98
Conference of Parties (COP), 58, 59, 107
congress, 16–19, 33, 35, 83, 99, 122, 144, 153
conservation culture, 144, 158
constitution, 17, 33
Convention on Biological Diversity, 58, 60, 63, 64
Convention on Certain Conventional Weapons, 64
Convention on the Banning of Biological Weapons, 64
Convention on the Banning of Chemical Weapons, 64
county, 14, 17, 19, 25, 26, 34, 40, 71, 83, 86, 145
criminal prosecution, 38
cultural revolution, 33, 54

Deng Xiaoping, 26
Denmark, 137

desertification, 56, 64
Designated National Authority (DNA), 138–140
Designated Operational Entities (DOEs), 139
developing countries, 22, 54, 60, 104–112, 114, 118, 122–126, 128, 132, 137, 141
development-first, 19, 22, 86, 115, 117, 118, 126
Dianchi Lake, 6, 76
diarrhea disease, 6
disarmament, 64
discharge fees, 18, 26, 34, 86, 156
discharge registration systems, 34
dispute settlement, 36
Dongtan, 121
drought, 10, 13, 103, 119
Du Shaozhong, 100

Earth Summit, 104, 106
Eastern Europe, 119
Eco-City, 120, 121, 149, 150
economic growth, 12, 22–24, 26, 27, 29, 31, 36, 40, 60–62, 79, 82, 105, 112, 115, 116, 118, 121, 123–125, 142, 144, 148, 150, 155, 156
economic structure, 77, 112, 114
eco-systems, 4, 15, 32, 76
Elizabeth C. Economy, 4, 7, 9
emissions trading, 114, 134, 135, 157
energy efficiency, 21, 31, 56, 113, 120, 125, 126, 136, 138, 149
enforcement of law, 7
environmental defense, 157
environmental diplomacy, 53–56, 58–61
environmental economics, 57
environmental engineering, 50, 96
environmental impact assessment, 34–36, 51, 56, 153
environmental law, 17, 33–39, 56, 57, 60
environmental NGO, 41–47, 49, 50, 157
Environmental Protection Bureaus (EPBs), 18, 19, 22–26, 34, 55, 73, 74, 81, 84, 89, 92, 96, 154
Environmental Protection Commission (EPC), 73
environmental tax, 86, 150, 151, 156
Euro I, 97
Euro III, 22, 118
Euro IV, 98
European Union, 8, 123, 135

farm forestry, 14
Finland, 137
fixed asset investment, 28, 112
flood, 1, 10, 103
foreign affairs, 4, 7, 9, 19, 21, 58, 116, 130, 140
foreign exchange reserves, 124, 132
foreign investment, 30, 31, 113, 128, 129, 137
fossil fuel, 11, 32, 62, 111, 113, 142, 151
Fourth Assessment Report (AR4), 121
Framework Convention on Climate Change (UNFCCC), 58, 60, 64, 104–107, 117, 125, 133, 134, 138, 139
France, 27, 49, 115, 129, 136
free-ride, 109
Friends of Nature, 41–43, 45, 46, 48
functional units, 18, 19

G-77, 105, 108, 124
G-8, 104, 116, 125
Gansu, 3, 145
General Agreement on Tariffs and Trade (GATT), 30
Geneva, 58
George W. Bush, 110
Germany, 27, 115, 116, 136
giant panda, 56
global commons, 56
Global Village of Beijing, 43–45
global warming, 11, 13, 60, 63, 99, 105, 106, 108, 109, 115, 116, 119, 124, 131, 141
globalization, 28–30
Goh Chok Tong, 149
government-organized NGOs, 42, 43
Great Hall of the People, 83
Green Camp, 45
Green Earth Volunteers, 43, 45
green GDP, 23
Green Olympics, 10, 88, 91, 97, 99, 158

Green Plateau, 45
green-credit policy, 146, 147
greenhouse gases (GHG), 10–12, 62, 64, 103, 104, 107, 108, 111, 118, 123, 127, 128, 130, 131, 135, 139, 141
gross domestic product (GDP), 6, 9, 11, 23, 26–28, 49, 78, 79, 81, 96, 112–116, 126, 137, 138, 142, 148, 154
groundwater, 4, 7, 13
growing pains, 77, 114
Guangdong, 3, 24, 79
Guangzhou City, 24
Guanting, 98
guanxi, 18
Guo Hu, 100

Haier, 98, 99
Haihe River, 2, 145
Han River, 1
Handan, 145
Harbin, 6, 66–70
hazardous waste, 13, 14, 58, 152
heavy industry, 11, 28, 79, 94, 112, 120
Hebei, 3, 94, 95, 99, 100, 145
Heilongjiang provinces, 6, 66, 68, 69, 75
Hejin, 145
Henan, 3, 145
Henry Paulson, 99
hot money, 63
Hu Jintao, 69, 116, 144
Huaihe River, 2, 35, 73, 77, 78, 145
Hubei province, 56, 151
human health, 10, 14, 92, 101
Hunan province, 56, 151
Huzhou, 79
hydrofluorocarbon (HFCs), 12, 99, 130
hydropower, 47, 117, 136, 145, 146

import quotas, 31
India, 12, 98, 105, 106, 108, 111, 121, 123–126, 128, 132, 134, 141
Indonesia, 12, 32, 121
industrial dust, 9
industrial revolution, 10
industrialized countries, 54, 55, 104, 106–108, 113, 114, 127
inefficiencies, 7
Inner Mongolia, 3, 145
institutional building, 77, 114
intellectual property rights, 64
Intergovernmental Negotiating Committee (INC), 105
Intergovernmental Panel on Climate Change (IPCC), 11, 12, 58, 62, 103, 121
Interim Regulations on the Prevention of Water Pollution in the Huai River Valley, 35
International Energy Agency (IEA), 11, 103
international environmental treaties, 21, 64
international institutions, 56, 62–65, 105, 109, 110
international norms, 54, 57, 60, 61, 63, 106, 109
International Olympic Committee, 90
International Union for Conservation of Nature (IUCN), 57
Internal Rate of Return (IRR), 135, 136
internet, 41, 43–45, 47–49, 146
Italy, 115, 136

Jacques Rogge, 90
Jakarta, 58
Japan, 10, 27, 49, 59, 69, 90, 107, 113, 115, 119, 125, 128, 130, 132, 136
Japanese Carbon Fund (JCF), 137
Jiamusi City, 68
Jiang Yanyong, 84
Jiang Zemin, 59
Jiangsu, 3, 23, 76–80, 131, 137
Jiaxing, 79
Jilin, 3, 6, 66–68, 70–72, 75
Jilin Petrochemical Company, 6, 67, 72
Joseph S. Nye, Jr, 61, 64, 110

Keppel Corporation, 150
Khabarovsk, 68
Kofi Annan, 70
Komsomolsk, 68
Kunming, 56

Kyoto Protocol, 10–12, 59, 60, 64, 103, 107–115, 121–125, 127–131, 134, 135, 139, 141, 143

landfills, 13, 14
Lanzhou, 145
Law on Conserving Energy, 35, 36
Law on Prevention and Control of Pollution from Environmental Noise, 35, 36
Law on Prevention and Control of Water Pollution, 35, 36, 39
Law on the Environmental Impact Assessment, 35, 36
Law on the Prevention and Control of Atmospheric Pollution, 35
Law on the Prevention and Control of Environmental Pollution by Solid Waste, 35, 36
Law on the Protection of Wildlife, 35, 36
Lee Hsien Loong, 150
legal liability, 35
letters and visits, 38
leukemia, 68
Lhasa, 56
Li Zhaoxing, 122
Liao, 2
light industry, 79
Lingbi County, 25
litigation, 36, 37, 74, 75, 83
Liu Jianchao, 122
Liu Qi, 49
London Conference, 58

Ma Jun, 1, 7
Ma Kai, 21
macro-control policies, 28, 112
Madrid Agreement, 64
Malaysia, 90
Manmohan Singh, 126
manufacturing sector, 31, 119
marine ecosystem, 4
market economy, 26, 28, 78, 155
Marrakech Accords, 135
mass media, 41, 46, 83
meteorology, 50
methane, 10, 117
Ministry of Agriculture (MoA), 19, 21, 38, 73, 140
Ministry of Civil Affairs, 49
Ministry of Commerce, 19, 129
Ministry of Construction, 19, 21, 73
Ministry of Environmental Protection, 19, 38, 39, 44, 77, 91, 100, 117, 153, 154
Ministry of Finance, 21, 120, 130, 136, 147, 149–151
Ministry of Foreign Affairs (MoFA), 19, 21, 58, 116, 130, 140
Ministry of Health, 21
Ministry of Land and Resources, 73
Ministry of Science and Technology (MoST), 21, 22, 117, 118, 130, 136, 140
Ministry of Water Resources, 2, 5, 21, 38, 73, 153
Montreal Protocol, 58, 59, 64
municipal treasury, 25
Myanmar, 32

Nairobi, 58
National Audit Office, 21
National Climate Change Program, 9, 12, 13, 62, 104, 115–117
National Coordination Committee on Climate Change, 21, 117, 118, 123, 130, 135, 136
National Development and Reform Commission (NDRC), 9, 12, 13, 19, 21, 22, 38, 62, 86, 104, 111, 116–118, 120, 130, 136, 139, 140, 146, 147, 149
national interest, 37, 53, 55, 62, 105
National People's Congress (NPC), 17, 19, 33, 83, 99, 122, 153
national security, 1, 37
national sovereignty, 22, 56, 60, 106, 118
national treasury, 25
Nations Conference on the Human Environment (UNCHE), 54–56, 60
natural and cultural heritage, 14
natural gas, 9, 12, 71, 98, 113, 114
negative externalities, 53, 109

Neil Carter, 24, 29, 104, 155, 157, 158
nepotism, 26
Netherlands Environmental Assessment Agency (MNP), 11, 62
netizens, 47
New York Academy of Sciences, 84
Ningxia, 3, 145
nitrogen dioxide (NO_2), 89, 91, 94
nitrogen oxides (NO_x), 94, 95
non-infringement of sovereignty, 53
non-interference of domestic affairs, 53
non-profit enterprises, 42, 43
Non-Proliferation Treaty, 63
North China, 1, 4, 99
North–South Water Project, 37
Northeast Asian Conference on Environmental Cooperation (NEAC), 59
Norway, 137
Nu River, 46, 47, 51
nuclear testing, 55

oceanic administration, 21
oil, 9, 32, 67, 113
Olympics, 10, 88–99, 101, 154, 158
opening-up, 26, 29, 33, 42, 54, 57, 66, 78
Organization for Economic Co-operation and Development (OECD), 7, 8, 13, 14, 113, 129, 137
overheating, 28, 112
overstaffing, 25, 26
ozone layer, 58, 59

Pacific Ocean, 69
Pan Yue, 44, 73, 152
Panjin Black-Beaked Gull Protection Association, 46
particulate matter (PM), 89–91, 95
pecking order, 17, 118
people's mediation, 37
perfluorocarbons (PFC), 12, 130, 131
Persistent Organic Pollutants (POPs), 59, 64
petroleum, 6, 12, 71, 113, 114, 124
plaintiffs, 37
political ideas, 61
political stability, 37
political systems, 16, 17, 41, 46, 77, 84, 86, 114, 155, 156
pollution fees, 24–26
PRC Environmental Protection Law, 33
premature death, 6
procuratorate, 18
Project Design Document (PDD), 139–141
Project Idea Note (PIN), 139
Prototype Carbon Fund, 137
provincial treasury, 25
public awareness, 48, 77, 96, 114, 129, 154, 158
public goods, 10
public interest, 74, 75, 158
Pudong, 151
Puyang, 145

Qinghai–Tibet plateau, 13, 119
Qu Geping, 55

raw materials, 28, 31, 32, 80, 112
Red tides, 4
reform, 7, 9, 12, 13, 19, 21, 24, 26, 30, 33, 38, 42, 54, 57, 62, 66, 78, 104, 111, 116, 117, 130, 139, 144, 146, 147, 151, 153, 158
Regulations on the Administration of Medical Wastes, 35
Regulations on the Control Over Dumping Wastes into the Sea Waters, 35
Resources for the Future, 157
Richard Louis Edmonds, 54, 57
Rio de Janeiro, 104
Rotterdam Convention, 64
Russia, 7, 32, 68, 69

SARS, 72, 84
Sasakawa Environment Prize, 70
Saudi Arabia, 124
Sea of Japan, 69
Sea of Okhotsk, 69
Security Council, 64, 105, 124
Shaanxi, 3, 56, 145
Shandong, 3, 79, 94, 95, 145
Shanghai, 3, 76, 77, 80, 119, 121, 131, 151
Shanxi, 3, 9, 94, 95, 145

Shenxian, 145
short messages, 48
Shougang Group, 99
Sichuan province, 56
silver iodide, 99
Singapore, 90, 120, 149, 150
social harmony, 37
social participation, 77, 114
soft power, 12, 53, 54, 61–65, 105, 106, 109, 110, 125, 126
soil erosion, 14
solar energy, 9, 117
solar power, 98, 113, 145, 146
Song Jian, 58, 73
Songhua, 2, 6, 66–75, 156
Songhua River, 6, 66–75, 156
Songliao Committee, 73
Songyuan, 68
soot, 9, 94
South America, 32
South Korea, 90, 125
Southeast Asia, 32
Soviet Union, 119
Spain, 137
species assessments, 57
spillover, 32, 97
stakeholder, 6, 39, 139
State Administration of Taxation, 150
State Commission on Environmental Protection (SCEP), 58
state council, 16, 17, 19, 21, 26, 35, 38, 47, 54, 73, 85, 114, 148, 150, 153
State Council Executive Conference, 21
State Council Information Office, 69
State Council Plenary Conference, 21
State Council Regulation on Pollutant Discharge Fee Collection and Use, 26
state councillors, 21, 73, 120
State Development and Planning Commission, 117
State Electricity Regulatory Commission (SERC), 145
State Environmental Protection Administration (SEPA), 2, 3, 5–7, 9, 19–22, 35, 44, 56, 59, 66, 69–73, 77, 78, 83, 84, 86, 91, 92, 96, 117, 118, 140, 145–147, 150, 152–154
state-centric, 16, 41, 80, 87, 114, 157
statistical life, 6
Stefanie Beyer, 33, 35, 37, 39
Stockhalm, 54, 59, 64
Stockhalm Convention, 59, 64
Strait of Tartary, 69
student organizations, 42
Su Wei, 122
subsidies, 30, 31, 129, 146
sulfur dioxide (SO_2), 8–10, 62, 89, 91, 94, 96, 101, 151, 154, 157
Sun Gang, 150
super-ministries reform, 153
Supervision and Management Agency for State Owned Assets, 148
surface water, 2–4, 7, 79
Suzhou, 79
Sweden, 137

Taihu Lake, 6, 16, 24, 76–78, 80–87
Taiwan (Republic of China), 49, 54, 60
Tang Jiaxuan, 120
Tenglong Aromatic PX Co., 48
terrorism, 63
The China Environment Yearbook, 4, 8, 9, 44–47, 50, 56, 58, 59, 67, 71
The China Quarterly, 14, 41, 43, 44, 46, 49
thermal power, 113, 117
third force, 42
Third World, 108, 124
Three Gorges Dam Project, 37
Tian Weiyong, 84
Tiananmen, 106, 109
Tianjin, 3, 7, 46, 94, 95, 100, 120, 121, 150, 151
Tianjin Friends of Green, 46
Tibetan Antelope Information Center, 46
tidal power, 145
Total Emissions Control (TEC), 157
township, 17, 19, 81, 85, 86
Township and Village Enterprises (TVEs), 81
trade barriers, 30
trade liberalization, 28–30
trade protectionism, 19, 63
traffic congestion, 13, 119
transaction costs, 63, 110, 139

transnational crimes, 63
tropical storms, 10, 103
two separate lines for revenues and expenses, 24, 25

Umbrella Carbon Facility, 137
United Kingdom, 12
United Nations (UN), 11, 54, 56, 58, 60, 63, 64, 70, 89, 104–107, 110, 111, 115, 121, 124, 125, 129, 132–136, 138, 140, 141
United Nations Conference on Environment and Development (UNCED), 60, 106
United Nations Convention to Combat Desertification, 64
United Nations Development Program (UNDP), 115, 136
United Nations Environment Programme (UNEP), 56, 58
United Nations Framework Convention on Climate Change (UNFCCC), 58, 64, 104–107, 117, 125, 133, 134, 138, 139
United States, 10–12, 27, 47, 54, 59, 62, 64, 84, 90, 99, 108, 110, 113, 119, 121, 123, 125–127, 141, 142, 157
Universal Copyright Convention, 64
upper-respiratory infections, 89
urbanization, 7, 27, 80, 101, 112, 155

Vienna Convention, 59
Vienna Meeting, 58
Vietnam War, 55

Wang Yubin, 99
Wang Yuqing, 71
water crisis, 1, 5, 69, 76, 80, 81, 85
water law, 6, 35, 36
water pollution, 1, 4, 6, 7, 24, 31, 35, 36, 39, 51, 56, 71–74, 85, 145, 153
water quality, 2, 4–6, 68, 71, 77–79, 84, 153, 154
water shortages, 7, 8
Weinan, 145
well-off society, 144
Wen Jiabao, 28, 47, 57, 85, 96, 112, 119, 120, 122, 147, 149, 150
West Asia, 32
wetlands, 14, 56
wildlife, 14, 35, 36
wind power, 9, 98, 113, 145
winner takes all, 28
World Bank, 2, 3, 5–7, 9, 56, 114, 115, 120, 128, 129, 131–133, 136–138, 149
World Health Organization (WHO), 89, 101
World Heritage, 57
World Trade Organization (WTO), 27–32, 64, 112
World Wide Fund for Nature (WWF), 56
world workshop, 28, 76
Wu Lihong, 82–84
Wuhan, 2, 56, 151, 152
Wuhu, 145
Wuxi, 24, 76, 79–81, 83, 85

Xi'an, 56
Xiamen City, 48
Xiangfen, 145
Xiangtan, 151
Xie Zhenhua, 6, 66, 70, 149
Xinhua News Agency, 4, 67
Xu Wen, 150

Yang Jiechi, 120
Yangtze River, 1, 4, 13, 56, 73, 119, 145, 151
Yellow River, 1, 8, 13, 119, 145
Yixing, 80, 83, 84
Yunnan Province, 47, 56

Zeng Peiyan, 119, 120
Zero O'clock Action, 77, 85, 86
Zhang Lijun, 69, 84
Zhejiang, 3, 76, 77, 79, 130
Zhou Shengxian, 70
Zhoukou, 145
Zhoutie Town, 81, 82, 84, 85
Zhu Rongji, 59
Zhu Tong, 96
Zhuzhou, 151